# THE SOCIAL GEOGRAPHIES OF MEXICO

# THE SOCIAL GEOGRAPHIES OF MEXICO

## Five Contemporary Case Studies of Society and Space

DAVID M. WALKER

THE GUILFORD PRESS
New York London

A Division of Guilford Publications, Inc.
*www.guilford.com*



Printed in the United States of America

This book is printed on acid-free paper.

For product and safety concerns within the EU, please contact *GPSR@taylorandfrancis.com*, Taylor & Francis Verlag GmbH, Kaufingerstraße 24, 80331 München, Germany.

Last digit is print number: 9 8 7 6 5 4 3 2 1

Library of Congress Cataloging-in-Publication Data is available from the publisher.

ISBN 978-1-4625-3386-2 (paperback)
ISBN 978-1-4625-5777-6 (cloth)

# Preface

The title of this book, *The Social Geographies of Mexico: Five Contemporary Case Studies of Society and Space*, requires a caveat. By no means is this a comprehensive analysis of Mexico's social geographies. Mexico is a large country, with nearly 130 million people, filled with millions of facts and figures, thousands of places, and lifetimes of experiences. This book is not intended as an exhaustive geographic survey of Mexico. The five stories about the social geographies of life in Mexico—Mexico City's historic center and Tepito neighborhood, the Costa Chica of Guerrero and Oaxaca, Tijuana, the garbage networks and dumps in Mexico City and its surroundings, and the social production of buffer villages in the Chimalapas—analyze the interrelationships between society, space, and place, and investigate how the nature and activity of these places are geographically created or changed by spatial patterns and processes as well as how these places are geographically expressed. This book poses questions about and analyzes how the social relations between society and space in these case studies are not fixed but are fluid and changing, how space plays an active role in establishing and producing social identities, and how social identities and relations produce material and symbolic spaces.

The chapters are interspersed with vignettes that provide definitions for theories and concepts and tell stories that will help the reader more deeply understand the places and peoples described in the case studies. Suggested additional readings, documentaries, and websites, when pertinent, are provided at the end of each chapter. After reading the book, my hope is that its approach and engagements, theories, and concepts will instill a sense of agency and possibility in readers, enabling them to imagine themselves asking questions about and analyzing socio-spatial phenomena, with similar or dissimilar complex and challenging stories.

# Contents

# CHAPTER 1

# Introduction

Mexico's landmass appears as a large, horn-shaped container or cornucopia. Mexico as cornucopia offers an abundance of diversity in its peoples and landscapes. The country is very urban. Its capital city is one of the largest urban centers in the world and a global city. The country also consists of the Chimalapas region in the in the Isthmus of Tehuantepec. The *selvas* of the Chimalapas contain some of the highest levels of biodiversity found in the world. Mexico is considered part of the Global South and, according to the World Bank, is now an upper-middle-income country. The country has the distinction of being the world's 14th-largest economy and shares a 2,000-mile-long border with the United States, the world's superpower both economically and militarily—a fact not left unnoticed by the turn-of-the-20th-century dictator Porfirio Diaz, who proclaimed, Poor Mexico, so far from God and so close to the United States. The population in the north of Mexico is more Mestizo, or of mixed Spanish and Indigenous heritage, whereas the central highland plateaus and the southern states of Oaxaca, Chiapas, Tabasco, and the Yucatan Peninsula are populated with dozens of Indigenous communities speaking distinctive languages. According to the 2020 census, 18.74 percent of Mexico City's population identified as Indigenous peoples who migrated to the capital in search of better economic opportunities, which further blurs the false dichotomies of rural = Indigenous and urban = mestizo. Black Mexicans, who refer to themselves as Afromexicanos or *Costeños*, predominantly live in communities in the Costa Chica of

Guerrero and Oaxaca and the Gulf State of Veracruz, consisting of more than 16 percent of the country's Afromexicano population. However, all 32 states have Afromexicano populations while 9 Mexican states have Afromexicano populations that exceed 2% and those 9 states account for 31.2% of the Afromexican citizens living throughout the country. Highland desserts, rich woodlands, vast coastlines and rich natural ports, large river systems and watersheds in the south of the country, and the Sierra Madre Occidental and Oriental mountain ranges create the extreme physical geographic features of the country. The social landscapes and geographies of Mexico are indeed rich.

## Chapter Outline

After indicating the important reasons to study Mexico, the next section describes the theoretical framework of the chapter: the social relations between society and space. The third section discusses Indigenous Mexico, including Indigenous civilizations and technology, and *Mestizaje*. In the fourth section, important moments in post-Revolution (ca. 1934) 20th- and 21st-century Mexican politics and political changes are delineated, from economic and political shifts, to Zapatismo, to political decentralization and the possible emergence of a codified multicultural Mexico. The fifth section provides an outline of the book and describes the research methods. The information provided in this chapter is intended to inform the reader about the larger structural settings wherein these case studies occur. The last two sections explain who this book is for and provide suggestions on how to use this book.

## Why Study Mexico?

This is a great moment to study Mexico and its people as the world (but especially the United States) needs to recognize the diversity, resilience, entrepreneurialism, and tenacity of the Mexican people in this increasingly interconnected global society. Mexico is too often commonly associated with tequila, beaches, fiestas, and cacti on one hand, and poverty, drug trafficking, violence, and workers migrating to the United States on the other. These binary perceptions do hold some degree of truth, but they fail to capture the ecological and cultural richness and global importance of the country. Indeed, *Los Estados Unidos de Mexico*, or the United States of Mexico, which is the country's official title, consists of much more.

Mexico is the 10th-most populated country in the world with 132 million people. It has the 13th-largest gross domestic product in the world and sixth largest landmass. Mexico is currently the largest goods-trading partner of the United States and Canada. U.S. exports of goods and services to Mexico supported 4.9 million jobs in the United States in 2015 according to data from the Woodrow Wilson Center's Mexico Institute. Approximately 36 million people either born in Mexico who have emigrated to the United States or are of Mexican origin currently live in Mexico's neighbor to the north, *Los Estados Unidos de America*, the United States of America. The country's diversity is seldom understood by foreigners traveling to Mexico for business or pleasure. Aside from the important cross-border, transnational contributions from institutions of higher education at the U.S.–Mexico border (such as the Institute for Regional Studies of the Californias, housed at San Diego State University, and El Colegio de La Frontera Norte situated in Baja California Norte), there are no Mexican textbooks translated into English and studied in university classrooms across the United States and Canada and in the rest of the English-speaking world. As described in the Preface, while this book does not seek to provide an encyclopedic overview of Mexico, it offers insights for readers to begin to develop a more nuanced understanding of the country.

## Social Relations, Society, and Space: What Is Social Geography?

In the introductory classroom, the elements of geography are divided between the natural sciences (physical geography) and the social sciences (human geography). The divisions are further layered into subdivisions. Although human geography is divided into the subfields of political, economic, urban, cultural, environmental, and historical geography, for example, physical geography subdisciplines are divided into weather and climate, geomorphology, water, rocks and minerals, and plants and animals. However, these divisions do not reflect the way the world around us works.

For example, in Chapter 4 on Tijuana, downtown Tijuana is socially reproduced beginning in the early years of the 2010s by young Tijuana residents employing and selling culture—the vernacular Nortec music, art, and food. This activity brings about the production of space and the creation of places imagined by and intended for Tijuana residents' consumption, instead of how the spaces had been utilized for U.S.-tourist vice consumption in brothels and in Mexican fantasy-heritage cantinas. The distinctive reproduction of space in downtown Tijuana demonstrates the co-constitution and codependence of social and economic processes. In the newly reconfigured spaces of downtown Tijuana, culture and economy are mutually constitutive.

Chapter 6, which describes the Indigenous Zoque production of space in creating the buffer villages to protect their territory and control the resources of the Chimalapas, provides an additional example of how these (sub) disciplines are often falsely reinforced in the classroom and are not representative of what occurs in the world around us. The Zoques reconfigured their positionality by conflating nature and society—embedding themselves in and as part of nature and the natural resources found in the Chimalapas to produce and control space in this locale. This example of the conflation of nature and society demonstrates how environmental and sociocultural processes are integrated, or to further elaborate, how physical and human geography are so deeply intertwined.

Social geography does not occupy a discrete intellectual space that can be sealed off from human or physical geography subdisciplines. Social geography is expansive and eclectic, but not inclusive, or just merely wide open and all-encompassing. Yet, the most vital aspect of social geography is that "understanding space is fundamental to understanding society (and vice versa) . . . more generally, social geographers focus on the implication of this relationship [society/space] for social identities, social reproduction, social inequalities and social justice" (The Newcastle Social Geographies Collective, 2021, p. 6).

What phenomena are at the core of social geography? What are some of the key themes in social geography, and how do they relate to the case studies in this book? Additionally, where do the society–space relations concerning social identities, social reproduction, social inequalities, and social justice take place? The following section engages with these questions.

In social geography the relation of power to place and space and who and what are excluded from place and space is the key focus of *power and exclusion*. To question and analyze sociospatial processes related to social identities, social reproduction, social inequalities, and social justice is to interrogate "who holds power, how [power and exclusion] shapes experiences of space and place, how this affects less powerful people in society and how social power is contested" (The Newcastle Social Geographies Collective, 2021, p. 19).

Power relations are often constructed along binaries: colonizer–colonized, heteronormative–LGBTQIAP+, white–Black/non-white as well as the societal normative opposites of age, religions, and disabilities that exist in a particular mainstream society. Examples in this book reveal that power and the indicators used to differentiate the powerful from the powerless are not fixed categories.

For example, Chapter 4, "The Social Geographies of Tijuana: Urbanization at the U.S.–Mexico Border," explains how landless migrants who came to Tijuana to work in the bourgeoning offshore assembly plants, called maquiladoras, created well-organized squatter organizations that mobilized large-scale land invasions. Once they became embedded in their newly constructed self-built homes, these squatter groups would politically negotiate with the landowners and authorities to earn titles to their plots of land. The powerless landless migrants reconfigured the power dynamic between the state/landowner and the landless laborer through these large and very public land grabs, demonstrating how power is continually shifting and can be resisted.

Another concern of social geography, intertwined with power and inclusion, is the theme of *justice and inequality*. Access to justice and equality hinges directly on access to and the ability to produce the spaces for justice and equality to occur. Access to health care, education, clean drinking water, food, shelter, and gainful employment are all very spatial endeavors. Each chapter in this book is concerned about access to justice and equality. The livelihood strategies of the street vendors and garbage pickers in Mexico City, the struggle for Afromexicano recognition and region formation to access development funds in the Costa Chica, the self-help housing in the irregular settlement juxtaposed to the reproduction and reappropriation of urban space in downtown Tijuana, and the Zoques fight for the production of lived places and space to protect their territorial heritage all center around spatial issues of justice and inequality.

The case studies analyzed in this chapter take place in specific communities and cities. Analyzing the processes that occur in *communities and cities*, another key theme in social geography, is aided by intertwining the subdisciplines of geography. The social production of space in Tepito, Tijuana, the historic center, and the Costa Chica requires analyzing the sociospatial, economic, cultural, and political practices and processes occurring in these locales. Indeed, to understand, for example, the cultures and economies of Tepito, or why Afromexicanos want to be recognized by the state as such, requires interrogating the social production of *identity and difference*, another key theme in social geography, occurring at Tepito and in the Costa Chica.

Social geographers (Panelli, 2004) describe how identity and difference may be actively conveyed or communicated, which can occur through cultural production and consumption or speech, such as in Tijuana with the production of Nortec music or in Tepito where the slang called *albures* is spoken. Identities can be read, recognized, and reinterpreted by others, such as the examples of race and ethnicity in the Costa Chica and the Chimalapas, respectively. Identities may be renegotiated, such as the change from *pardo* or *negro* to Afromexicano in the Costa Chica. Identities may be represented and performed, as the case of adorning the body with *La Santa Muerte* tattoos by Tepito residents or the choice of Afromexicanos to wear their hair in a natural style or braided in the Costa Chica. Finally, identities can take up, employ, or even reconstruct space, as the examples of street vending and public *Santa Muerte* worship do in Tepito and in the historic center of Mexico City.

To study where society–space relations concerning social identities, social reproduction, social inequalities, and social justice take place does not require only engaging these processes at the everyday scale, or on the ground. In contrast, to analyze, for example, the case studies found in this book requires engaging in the multiple scaling of their sociospatial phenomena, or the socially produced scales from the *intimate to the global*. Nonetheless, socially produced scales do not just concern size (global/local), level (municipal/federal), and relation (corporal/institutional) as scale is not "necessarily a preordained hierarchical framework for ordering the world–local, regional, national and global. . . . [Scale] is instead a contingent outcome of the tensions that exist between structural forces and the practices of human agents . . . scale is constituted and reconstituted around relations of capitalist production, social production and consumption, and that attention to all three sets of relations is critical to understanding fully the social construction of scale" (Marston, 2000, pp. 220–221). This book analyzes the social processes producing the distinct spaces discussed in each chapter through a multiscalar lens. For example, cross-sectional policies, practices, and relationships are found in Chapter 3 about identity and region formation in the Costa Chica, where the reconfiguration of geographic identity, that of being from the coast to demanding to be recognized by a racial categorization, as Afromexicano, is occurring at multiple sites. From the coastal villages to the municipal seats of Pinotepa Nacional and Cuajinicuilapa, to the capitol cities of Oaxaca and Guerrero states, to the seat of the federal government in Mexico City, Afromexicanos have lobbied to be counted as Black Mexicans in order to count as Mexicans. And through their struggles—as the production of scale is the result of contestation—they have created Afromexicano NGOs by working across scales with Indigenous civil society groups and other national and international Afro-heritage NGOs, effectively producing a spatial scale (the Afromexicano NGOs), where they can mobilize their power. Additionally, one of the goals of this struggle is to have the Costa Chica recognized by the state as an autonomous region; thus, if successful, they would effectively be producing a new scale to perform their cultural, economic, and social identities.

In this section I have provided the theoretical framework that guides how I analyze the case studies found in the book. Using this social geographic framework, this section has shown how social geography encompasses an expression of many of the connections in the subfield of geography. Employing this broad understanding of social geography reflects how I imagine the best way to analyze the world.

## Indigenous Mexico

The first known remnants of human life in Mexico are community fires found in the Valley of Mexico dating from 21,000 years ago. Indigenous peoples of Mexico domesticated corn from a grass called *teocintle* to selectively breed maize plants approximately 10,000 years ago. Maze became the most important plant in Mexico, which diffused across the Americas, and is the anchor of many origin narratives for the land's peoples. Until today, Mexicans conceptualize maize as not just a crop, but a deep cultural symbol intrinsic to daily life.

Evidence shows a marked increase in pottery working by 5,500 B.C.E. and the beginning of intensive corn farming between 4,000 and 3,700 B.C.E. Thus, Mesoamerica is one of the major cultural hearths of the world, emerging in tandem at roughly the same time as the Fertile Crescent of Mesopotamia and the Nile Valley of Egypt, based on wheat; the Indus River and Ganges River valleys of India and the Yellow and

Yangtze Rivers of China, based on millet and rice; and the Andes of Peru, based on the potato.

## Indigenous Civilizations

Four thousand years before contact with Europeans, Indigenous communities constructed complex cultures that matured into advanced civilizations. With advances in agricultural technology, such as the domestication of *teozintle*, and the intensification of maize, squash, bean, tomato, and cotton farming, surplus agricultural production coincided with the emergence of sedentary populations and the growth of villages and eventually cities.

The Olmecs (3,500 B.C.E.), the first major Meso-American civilization, produced pottery, fine art, and graphic symbols with which they recorded their history, society, and culture, as evidenced at the archeological sites of San Lorenzo and La Venta. A number of regional Indigenous communities eclipsed Olmec hegemony, including the Maya, Zapotec, Totonac, and Teotihuacan civilizations, who shared a common Olmec heritage. As the dominant Maya and Teotihuacan began to wane, the Toltecs came to dominate Meso-America until they too would be displaced by the nomadic tribes from the northern frontier. In 1325 the nomadic Chichimeca tribe of the Mexica—more commonly known as the Aztecs—formed a settlement called Tenochtitlán that would grow to be the well-built and organized city that amazed the Spaniards upon the arrival of Cortez and his entourage in 1519. They were awestricken when first laying eyes on the causeways crossing Lake Texcoco surrounded by the floating gardens called Chinampas that led into the largest city with a population 140,000, that they had ever witnessed. Tenochtitlan's cleanliness and order, replete with running water fountains and sewage culverts, was unparalleled compared to cities in the Iberian Peninsula (see Díaz del Castillo and Cohen, 1963, in "Suggested Resources")

## Indigenous Technology

Indigenous communities consisted of advanced societies. The Olmecs, especially, excelled in mathematics. They developed the concept of zero as early as 36 B.C.E. and worked with sums of hundreds of millions. In astronomy, the Olmecs, followed by the Maya, measured the length of a solar year to a high degree of accuracy. Across Meso-America pre-Colombian societies domesticated crops and developed irrigation, and the methods that led to the formation of the floating gardens called Chinampas (see Díaz del Castillo and Cohen, 1963, in "Suggested Resources"), which fed Tenochtitlan's residents. These pre-Colombian societies from the Toltecs to the Teotihuacan, Maya, Zapotecs, and Aztecs built pyramids, temples, and well-organized cities replete with running water and sewage.

## *Mestizaje*

I emphasize the position of Indigenous cultures in Mexico because they continue to be important today. In the first century after conquest, 90 percent of the 25 million people across Mexico died (see Crosby, 1972, and Mann, 2005, in "Suggested Resources"). A demographic rebound would occur in the late 18th century. Today more than 12 million people identify as Indigenous, and 6.2 million speak an Indigenous language as their maternal tongue (Instituto Nacional de Estadística y Geografía INEGI, 2020). Mexican Indigenous cultures such as the Aztecs in the Valley of Mexico, the Zapotecs and Mixtecs in Oaxaca, the Totonac in Veracruz, and the Maya in the Yucatan Peninsula and Chiapas are part of contemporary Mexican culture. The General Law on the Language Rights of Indigenous Peoples, which states that "indigenous languages will be valid, as well as Spanish, for any matter or process of public nature, as well as for full access to the management, services and public information" (Global Regulation, 2022), demonstrates the burgeoning political power that Indigenous groups have harnessed since the 1994 Zapatista Revolt.

Indeed, the national identity of Mexico is infused with Indigenous symbolism, from the eagle with a snake in its talons atop a prickly pear cactus—which is found on the Mexican flag—to the historical figures found on much of the country's currency, to place names and vocabulary.

Mexican foods are also highly influenced by Indigenous cuisines. Corn, chilis, tomatoes, chocolate, and agave for distilling mezcal and tequila or fermenting pulque; and avocados, squash, and beans—all crops endemic to Mexico—make up the core of Mexican cuisine to this day. Mexican foods also represent a form of cultural syncretism representing the fusion of Indigenous and Iberian ingredients and techniques, such as mole, the savory chocolate, chili, and dried-fruit sauce. The crops that emerged from Mexico inform how the world eats. Mexico is a mega-diverse country. As such, many ingredients consumed by people world-wide originate in Mexico as evidenced by their Nahuatl names. Globally, quotidianly consumed foods such as chocolate, chilis, tomatoes, maize, *camote*/sweet potatoes, vanilla, avocado, guava, *chayote*, *epazote*, *jicama*, for example, and many varieties of beans are used across the world only after the (dis)encounter that occurred with conquest, colonization, and global trade. Can you imagine Italy with no tomatoes, Ireland with no potatoes, Switzerland with no chocolate, and South Asian, Thai, and Indonesian cuisines with no capsicum/chilis? Or Mexico with no pork, for that matter? (see Crosby, 1972, in "Suggested Resources").

The capital of the nation itself is superimposed onto the Aztec city of Tenochtitlan, demonstrating how Mexican society has always already been urban. The Spanish retained the original layout of the city of Tenochtitlan, reflected today in the various city districts and in the central precinct of the Zocalo, or main square, which is located at the same site as the Aztec pyramid-temple. The power center of the Aztecs was much too important for the Spaniards to consider relocating. Many of the historic buildings in the *Centro Histórico* are built with the volcanic red rock called *tezontle* taken from the pyramids. Several pyramids and ruins remain unearthed within the urban sprawl of the city. Following independence and after a number of floods, the lakes of the valley were drained, drastically changing the landscape. The former island city now was able to spread over a dry plain. Only small remnants of the old canal city remain, such as in the celebrated flower district of Xochimilco.

## Indigenous Identity and *Mestizaje*

Overall, Mexico's identity is closely tied to its reimagined Indigenous past juxtaposed to its current, or living, Indigenous population. Mexicans do not consider themselves transplanted Europeans as do its two neighbors to the north. From Miguel Hidalgo's *grito* (cry) of independence in 1810 though the 19th century, which resulted in losing more than half the country's territory to the United States (in the War of the Northern Invasion 1848, as it is referenced in Mexican history books) and through the tumultuous Mexican Revolution (1910–1934), Mexicans began to form their identity as mestizos. This identity became codified by the federal government's implementation of education programs that emphasized the *mestizaje* (miscegenation) of the Mexican population. Led by José Vasconcelos as the minister of education, the educational doctrine celebrated the Indigenous past and vilified the Spanish as invaders (see Joseph and Henderson, 2002, in "Suggested Resources"). Vasconcelos's theory, which is described in detail in Chapter 3, promulgated the doctrine of *la raza cósmica*—the cosmic race—the combination of Indigenous people and Europeans to form the mestizo race. However, Vasconcelos's theory did not celebrate the living Mexicans of his time. He romanticized the technologically advanced Aztecs and Mayas and ignored the lack of services and education provided to rural Indigenous communities after the revolution. He also excluded what the Mexican scholar Guillermo Bonfil Batalla would eventually call *la tercera raíz,* the third root. The third root that forms *la raza cósmica* comes from Africa; thus *mestizaje* consists of the mixture of European, Indigenous, and people of African ancestry.

Although Mexico leans heavily on its Indigenous past to create a sense of nationalism, its contemporary Indigenous identity is more complicated. While there are changes afoot in the country's attitude toward its Indigenous population, *indio,* or Indian, remains a pejorative term. Changes in Mexico's census, the formation of autonomous Indigenous communities, and laws promulgating Indigenous linguistic rights may begin to change the nation's discriminatory view of its Indigenous and Black populations.

## 20th- and 21st-Century Mexico

The case studies discussed in this book do not exist within a vacuum. Decisions made by the three branches of government impact local economies and how people in these locales negotiate their lives. It is important to draw attention to the Institutional Revolutionary Party (PRI) that emerged from the Mexican Revolution and would rule the country from 1934–2000; the shift from a state-led economic policy to free-market neoliberalism from 1980–2020; the 1994 Zapatista uprising and its impacts on Indigenous identity and autonomy, which would also influence Afromexicano struggles to be recognized and counted by the Mexican state; and the Mexican state slowly recognizing and codifying the country's multiculturalism, which coincided with the end of PRI hegemony. This chapter does not cover these topics in depth (see "Suggested Resources" at the end of chapter for more comprehensive discussions on Zaptismo from Barry, 1999, and neoliberalism from Harvey, 2005). Nevertheless, it is important to outline these political, economic, and social trends and shifts in Mexico to help the reader understand the case studies.

## La Dictadura Perfecta

The goal of the Mexican Revolution was to overthrow the dictatorship of Porfirio Diaz. This singular goal would completely change Mexican society. The colluding factions that led to the ousting of Diaz had differing reasons for doing so and did not always join together with an ultimate objective. There was no set ideology on what to do next. The violence of 1910 gave a clear start to the Mexican Revolution, but it is more difficult to identify an end point. Many scholars suggest the year 1920, but some end it with the 1917 constitution or events in the 1920s. Others argue that the revolution slowly unraveled until 1940 (see Beezley and Meyer, 2010, in "Suggested Resources").

After the rise to, and fall from power of, six Mexican presidents between 1917 and 1934, Lazaro Cardenas would be elected for a single 6-year term and could not be reelected, a term agreement that exists until today. He changed the name of the ruling party to the PRI—the Institutional Revolutionary Party—distributed power among peasants, labor, the military, and the popular sector consisting of teachers and bureaucrats, and consolidated the power structure of the PRI for the next 70 years. His administration implemented land reform through the redistribution of large landed estates into *ejidos*, which are lands communally held in the traditional Indian system of land tenure that were distributed to landless peasants after the revolution and well into the 1970s. He nationalized the oil industries, paying the mostly British and some American companies that had invested in the extraction and refining of petroleum. The PRI consolidated power from the top down, thus effectively implementing economic policies; controlling labor; controlling foreign direct investment; and mandating the budgets of the state and municipal governments. This top-down, three-tiered Mexican federal/state/municipal system was based upon loyalty to the PRI. The PRI created a form of corporatist and clientelist governance that rewarded those who supported the PRI hegemony and punished those who opposed the PRI machine.

Under this system, if a municipal government wants anything from a state government, it must do what the state requests, on up the chain to the federal level; thus *La ley de Herodes*, or "Herod's Law," a Mexican saying that emerged from the resulting politics of the Mexican Revolution and also the title of a film satirizing political corruption, becomes the de facto mechanisms of power. *Te tocó la ley de Herodes, o te chingas o te jodes*—an inverted form of you scratch my back and I'll scratch yours (literally translated, means either you screw others over, or you get screwed, which means you have to do something you don't want but are obliged to do so and you must do so for your own good). Mexico held elections every 6 years from 1934–2000. Every 6 years a chosen PRI candidate—who had been given the thumbs-up—*el dedazo*—would win. Thus, the PRI system was declared the perfect dictatorship of Latin American. No coup d'etats, no military interventions and overthrows, and no direct military interference by the United States throughout the post-Revolution 20th century.

## Shifts from PRI Economics to Neoliberalism

From the 1940s through the late 1970s Mexico's economy and population grew rapidly. The PRI-controlled, state-owned, and mixed-capital enterprises, combined with a highly regulated private sector, consisted of oil exports, agricultural production, the building of factories for domestic consumption—under the plan called import substitution industrialization (ISI)—and a rapidly growing tourism sector. The government strictly controlled foreign investment and imports and barred private investors from ownership in many industries, including mining, forestry, insurance, and energy production. Semiautonomous state corporations managed all primary, secondary, and tertiary sectors of the economy. Petróleos Mexicanos (PEMEX), Cementos Mexicanos (CEMEX), Teléfonos de México (TELMEX), Comisión Federal de Electricidad (CFE), Banco Nacional de México (BANAMEX), and Ferrocarriles Nacionales de México (FNP) managed, generated, and distributed energy and cement, and controlled telecommunications, banking, railways, and airlines. In addition, the government regulated the prices of many goods and services.

This era also coincides with the Bracero Program in which guest workers from Mexico were invited to the United States to fill the labor shortage caused by World War II. The Bracero Program lasted from 1941 until 1964, when it was halted—not by the United States, but by the insistence of the Mexican government, who needed the labor to fuel the ISI occurring at the time. Four and a half million Mexicans participated in the Bracero Program. These laborers sent millions of dollars home as remittances to their families. It is important to emphasize this program as an example of how emigration from Mexico to the United States functioned as an important social escape valve and marked the initiation of the country's reliance on remittances. Today remittances are Mexico's third-largest source of income after oil exports and tourism.

The Mexican Miracle refers to this period of rapid growth under ISI, which entailed remittances, tourism, oil exports, agricultural production, and urbanization. It is an example of the success of state-led capitalism in comparison to other Latin American nations.

In late 1976 massive oil reserves were found in the Gulf of Mexico at the Cantarell Field, which from 1981–2004 was Mexico's largest oil field and the second largest globally. The time of discovery coincided with high oil prices provoked by a strong and unified OPEC, who benefited from lowering production to drive up prices. U.S. banks were eager to lend money and provide technological services to Mexico to exploit the newly discovered fields. The Mexican government borrowed heavily.

In the late 1970s the country was investing in programs for food self-sufficiency, education, infrastructure, and other social services, and building the infrastructure needed to exploit the Cantarell oil reserves (see La Botz, 1995, in "Suggested Resources"). In 1980 the United States shifted politically, electing a man who ran for president by debasing big government. His ideology was based on Thatcherism (the first neoliberal experiment in the Global North), with a healthy fear of communism and a firm belief in a strong military—ironically funded by big government. After the election, Reagan quickly convinced OPEC to increase production, which lowered oil prices. Between 1977 and 1981, PEMEX oil production doubled, but with lower oil prices, overspending, and rampant inflation the country was not able to finance the loans it had taken from U.S. banks. Furthermore, in 1982, in attempts to slow down global inflation, the United States increased interest rates. Highly indebted Mexico, which invested heavily to develop its oil field, was thrown into crisis. Interest rates increased dramatically at the same time that the demand for oil and prices decreased. In 1982, Mexico defaulted on its interest payments, being the first sovereign nation to do so.

Mexico had few options for what to do. It negotiated an International Monetary Fund (IMF) loan of more than $3 billion that required the country to implement structural adjustment policies, or SAPs. SAPS directed the government to implement a neoliberal economic policy, including completely deregulating many industries; dismantling, privatizing, and selling off many state enterprises; welcoming large amounts of foreign investments; removing most import restrictions; devaluing the Mexican peso; and increasing interest rates (a full definition of neoliberalism is provided in Chapter 2) Thus, the Mexican Miracle abruptly shifted to a neoliberal model. This shift greatly impacted the informal market, including street vending and garbage collection. As discussed in Chapter 2 on La Santa Muerte and in Chapter 5 on garbage, the informal market and street vending have existed in Mexico since Mexico City was called Tenochtitlan. The end of the Mexican Miracle and the shift to a lean, neoliberal model merely accelerated the informal economy's growth. This is further described in Chapter 2 as roll-back and roll-out neoliberalism. The retrenchment of the state from 1982 onward greatly impacted Mexican society, politics, and economics.

## Zapatismo

On the first day of January 1994 the *Ejercito Zapatista de la Liberacion Nacional* (EZLN), or Zapatistas, composed mainly of Mayan Indians, burst onto the world scene when it occupied seven towns in the Mexican state of Chiapas, among them San Cristobal de las Casas, the second largest town in the state. Contesting the neoliberal program and producing autonomous Indigenous spaces has always been at the heart of the Zapatista movement. The neoliberal program in Mexico, coinciding with the rollback of state supports, created spaces for social movements to contest power from below that did not exist during the reign of the PRI. This uprising came as a shock to the government of Mexico, which had expected the day to mark Mexico's installation into the First World at the outset of the North American Free Trade Agreement, or NAFTA, now called the United States–Mexico–Canada Agreement, USMCA (see Barry, 1999, and Hayden, 2003, in " Suggested Resources"). The Mexican government quickly moved to squash the rebellion, sending troops and bomber jets to Chiapas. However, under the guidance of Subcomandante Marcos, the Zapatistas ingeniously weaponized the internet, thus globally broadcasting the rebellion and converting the Zapatista revolt from a group of Mayans resisting the state's neoliberal model to a transnational movement. The eyes of the world were on Mexico in January 1994, and the state did not implement its hegemonic plans and obliterate the Zapatistas. They are still in the jungle today.

The weakening of the PRI's corporatists and clientelist mechanisms allowed groups like the Zapatistas to assert rights and demand autonomy based on ethnic *identity*, not as laborers and peasants. The Zapatistas successfully negotiated with the state to recognize Indigenous identity—not in the form of *mestizaje*—and form autonomous spaces, which would eventually become autonomous Indigenous municipalities. The goals achieved by the Zapatistas inform the case studies on Los Chimalapas, using Indigenous identity to create Indigenous municipalities governed by *usos y costumbres* (traditional Indigenous and land management agreements or laws), and the Costa Chica, the mobilization of Black identity, and the region formation found in this book.

## Multicultural Mexico?

Elections held in 2000 marked the first time since the 1910 Mexican Revolution that an opposition candidate, Vicente Fox of the National Action Party, or PAN, defeated the PRI candidate. He was succeeded in 2006 by another PAN candidate, Felipe Calderon, but Enrique Peña Nieto regained the presidency for the PRI in 2012. Then in December 2018, left-leaning, antiestablishment politician and former mayor of Mexico City (2000–2005) Andrés Manuel López Obrador, from the National Regeneration Movement or MORENA, won the election in a landslide and became president.

The devolution of the PRI machine has coincided with changes in Mexican society. Indigenous autonomous municipalities are recognized, and a law has been passed to protect "the linguistic, individual and collective rights of Indigenous peoples and communities, as well as the promotion of the daily use and development of Indigenous languages, under a context of respect for their rights" (Global Regulation, 2022). The protection of the LGBTQ community is being debated, and women's reproductive rights, including women's right to choose, are recognized, which represents impressive advancements in a country that is still 75 percent Roman Catholic. Coinciding with these changes are the thousands of Mexicans who have chosen to leave the United States over the past decade and return home. These returning migrants decided to go back to Mexico for a variety of reasons. As they return, they bring with them distinct views of society, especially in relation to women's and LGBTQ rights. They are also using skills developed in the United States and investing in Mexico.

## Methods and Book Outline

To learn about the formation and production of the places described in this book required a comprehensive approach. For each of these case studies I conducted open-ended interviews, structured interviews, participant observation, archival research, and, in the case study of the Costa Chica, employed visual methods. Triangulating these methods provides a fluent approach for learning about these distinct places. Additionally, conducting the research at the sites found in these five case studies called for embedding myself in and among the people who are producing and being shaped by these places, which involved engaging with as many actors as possible at each site.

### *Chapter 2: Shrines of the Times: The Social Geographies of Contested Space in Tepito and the Historic Center of Mexico City*

This chapter analyzes the corollary links among shifting economic policies in Mexico, the proliferation of the informal economy, and the intensification of Santa Muerte adoration in Mexico City's historic center and the Tepito neighborhood. Within this context, a description of Mexico City's neoliberal public–private revitalization plan of its historic center, called *El Programa de Rescate* (recovery program), is described, focusing on stakeholders' interest in the production and utilization of urban space.

I have been gathering documentation on Mexico City from 1989 until the present. More specifically, I spent a year living in Mexico City doing research for my dissertation. My mixed-methods approach includes doing archival research, conducting participant observations, interviewing the director of the Center for Tepito Studies, as well as street vendors, market workers, Santa Muerte worshipers, Doña Queta, who cares for the Santa Muerte Shrine in Tepito, and business owners. I volunteered at the Center for Tepito Studies and Grupo Capitalista, an informal street-vending organization. I made connections at UNAM (the Autonomous University of Mexico) where I volunteered to teach urban geography for one semester, which provided an institutional affiliation that helped opened doors for me to interview city officials and attend conferences on land use in Mexico City.

### *Chapter 3: Identity and Place: The Social Geographies in the Costa Chica of Guerrero and Oaxaca*

This chapter tells the story of the people of African descent who live in the coastal region of southern México, and how Afromexicanos produced the social geographies found along the Costa Chica. Currently, Costa Chica residents (*Costeños*) are mobilizing their Black political identity and consciousness as it relates to the production of region and place to make formal demands for an official regional recognition of the Costa Chica by the national and state Mexican governments with the goal of receiving federal and state development aid and of controlling natural resources.

I first visited the Costa Chica while living in Oaxaca City in 2004. I returned to conduct research for 2 months in the summer of 2012, a month in December–January of 2012–2013, and spent 2 months in the region in the summer of 2013. I returned for a 2-week visit in the summer of 2017. I stayed with Valentin's (a resident fisherman and friend) extended family in Corralero and at small hotels in the regional capitols of Pinotepa National, Oaxaca, and Cuajinicuilapa, in Guerrero.

Valentin and his family live in a large compound made of cinderblock homes surrounding a courtyard filled with palm and mango trees. Here I slept in a hammock and spent time with Valentin and his family gathering clams in the lagoon, socializing at the compound, and going out in the open Pacific Ocean in his 18-foot fiberglass boat only equipped with an outboard motor. No GPS, no radar, no radio. We'd leave at 3:00 A.M. and return around 11:00 A.M. with 30- to 50-pound tunas or a 10- to 15-pound snapper. Valentin would sell this haul to the local restaurant or drive it into Pinotepa—only about 15 miles away, but an hour drive. Without refrigeration, he could not preserve the fish and get it to markets in Acapulco or even Mexico City, where he could demand more money for the fish.

I incorporated participant observation at Valentin's compound, at the Museo de las Culturas Afro-Mestizas de la Costa Chica in the town of Cuajinicuilapa, and at La Casa de La Cultura Afro-Mestiza de La Costa Chica in the city of Pinotepa Nacional. I conducted structured interviews with the director of La Casa de La Cultura Afro-Mestiza and unstructured interviews with Valentin and his family and many others, and did archival research and visual content analysis at the museum.

### *Chapter 4: The Social Geographies of Tijuana: Urbanization at the U.S.–Mexico Border*

This chapter explores the social production of Tijuana: how social, political, and physical geography have shaped the economies, cultures, and urban morphology of the city, as well as how the spaces of the city impact its residents and the Tijuana–San Diego border region. Geography matters. This case study centers on analyses of government policies from both sides of the border combined with the stories of migrants, artists, and entrepreneurs and how the intertwining transnational processes make the city what it is today.

I have been researching and visiting Tijuana since I was an undergraduate student in northern California. I lived in the San Diego–Tijuana region from 1996–2001, earning a master's (MA) degree and working in Tijuana. My MA thesis focused on the privatization and urbanization of *ejidos* in Tijuana. Rents were too high for a graduate student, so I rented an apartment in Tijuana and traveled back and forth across the border. I have continued to visit Tijuana at least once every year until 2017. COVID has interrupted many of my travel plans. I owe myself a trip. The research for this chapter includes years of archival research, interviews, and participant observation.

### *Chapter 5: In the Dumps: The Social Geographies of Trash*

This chapter centers on the politics of and livelihood strategies and struggles over trash. It analyzes the shifting role that garbage plays in the urbanization of Mexico City and the garbaged networks of relationships between formal and informal economic activities and political–client patron relations among *pepenadores*/garbage pickers and *caciques*/political organizers. As the country's political and economic structural change that began in the 1980s continues with the maturation of neoliberal policies well into the 21st century, garbage too is impacted by disciplinary neoliberal doctrine, which affects the social politics in this locale. Population and waste production increases are linked to the story of garbage sorting in and around Mexico City as a metaphor for the shifting political economy of the city and country: from clientelist PRI state and *caciquism* to privatization as part of the neoliberalization of society.

The research for this chapter is based on archival fieldwork, participant observations, and open-ended interviews conducted in Mexico City and Alpuyeca, Morelos. In the summer of 2017, I earned a scholarship to study at the Cemanahuac Institute in Cuernavaca, Morelos. I used the time to conduct my research into the garbaged networks of Mexican trash.

### *Chapter 6: Zoque Indigenous-Produced Space: Social Geographies of Los Chimalapas*

This chapter describes how Indigenous Zoques implemented material and metaphorical practices that are embedded in a larger framework of conceptualizations of space and "nature" into the social production of buffer villages to prevent the encroachment of illegal loggers, landless peasants, and NGOs.

Archival research, open-ended and structured interviews, and participant observations were used to gather the information for this chapter. I lived in Oaxaca for 1 year as a research assistant working on a grant studying the interconnectivity of international NGOs, national NGOs, local NGOs, and the communities on the ground where these NGOs were located. I spent the year volunteering at the World Wildlife Foundation regional offices of the World Wildlife Foundation (WWF) in Oaxaca. I also worked on the biosphere project located in the Chimalapas. During my time there I traveled to the villages of San Miguel and Santa Maria on four separate occasions as a volunteer for the WWF. I also had three random opportunities to meet and talk with Zoque leaders without the supervision of WWF officials when they showed up at the WWF offices unannounced and only I and the secretaries were present. I invited them to a nearby cantina to win their confidence and learn how they conceptualized their relationship to place and space in the Chimalapas. And on one occasion I met with Zoque leaders in the regional city of Juchitan to discuss the Zoque strategy of the production of buffer villages along the border of their territory.

## Whom Is This Book For?

*The Social Geographies of Mexico* will appeal to instructors, undergraduate students, and beginning MA and PhD students in geography, Latin American and Latino studies, cultural studies, and across the social sciences, who want to learn more about the production of place in Mexico. In K–12 public school classrooms, we learn so little about the peoples and places of Mexico. This book describes in detail the production of place and space in the locales highlighted in each case study. In doing so, it will help reduce this knowledge deficit among the increasingly multicultural audiences found at 21st-century college campuses. So, while this book does not aim to be encyclopedic, it does provide an entry point for learning about life in these places.

## How to Use This Book

Although this book is not a geographic survey of Mexico that systematically describes the peoples and cultures, economic development, physical features, and environmental issues found across the entirety of the country, it can be used on its own as a core textbook to teach students about the social production of place and space, and about how identity influences and is influenced by the geographic sociospatial dialectic in these locales. Each of these engaging stories introduces geographic theories and concepts as they discuss how the social geographies of these places are produced and imagined. The introduction, which can be elaborated on by the instructor, informs the reader about the larger structural settings wherein these case studies occur. Although the chapters can be read in the order in which they are presented, there is no reason why the order cannot be switched according to each instructor's plan.

This book can also be used as a supplemental text for survey courses on Mexico and Latin America. In this case, the chapters could be chosen by the instructor to provide an in-depth analysis of place beyond the usual understanding of facts and figures. For survey courses, any of the chapters could be assigned to provide an insight into how people are negotiating and reconfiguring their lives in distinct locales across Mexico. My hope is that instructors can add many additional, similar engagements with Mexico from their own experience and the experience of others, and together build a sense of the importance of deep engagement with a place and its people.

This book can also be used as a supplement to other materials. For example, instructors teaching courses about race in Latin America could assign Chapter 3, "The Social Geographies of the Costa Chica," to provide an example of how Blackness is being reimagined across Latin America and Mexico. Instructors interested in NGOs, or indigeneity, can assign Chapter 6, "The Social Geographies of Los Chimalapas," as supplemental reading that provides a case study of Indigenous professionalization and the shortsightedness of NGOs working in Mexico and throughout the region. A political science instructor on Mexico could assign the chapter on garbage as an example of clientelism and its shifts under the devolution of the PRI.

The book would also be useful for K–12 teachers. Here it would function as a professional development tool to help instructors gain a well-rounded understanding about the places from where their students and students' families emigrated.

## SUGGESTED RESOURCES

To learn more about the NAFTA's impacts on corn production and the Zapatista uprising, see:

Barry, T. (1999). *Zapata's revenge: Free trade and the farm crisis in Mexico*. Boston: South End Press.

Hayden, T. (2003). *The Zapatista reader.* New York: Thundermouth Press/Nations Books.

Joseph, G. M., and Henderson, T. J. (2002). *The Mexican reader: History, culture, politics*. Durham, NC: Duke University Press.

La Botz, D. (1995). *Democracy in Mexico: Peasant rebellion and political reform*. Boston: South End Press.

For a concise history of Mexico for those who would appreciate more historical context, see:

Beezley, W. H., and Meyer, M. C. (2010). *The Oxford history of Mexico*. New York: Oxford University Press.

To learn more about pre-Columbian Indigenous society, see:

Berdan, F. F., and Smith, M. E. (2020). *Everyday life in the Aztec world*. New York: Cambridge University Press.

Coe, M. D. (2022). *The Maya*, tenth edition. New York: Thames & Hudson.

Coe, M. D., Urcid, J., and Koontz, R. (2019). *Mexico: From the Olmecs to the Aztecs*, eighth edition. London: Thames & Hudson.

Flannery, K. V., and Marcus, J. (2003). *The cloud people: Divergent evolution of the Zapotec and Mixtec civilizations*. Clinton Corners, NY: Percheron Press.

Roys, R. R. (1933). *The book of Chilam Balam of Chumayel: Literature of the Yucatan Mayans; The religion, calendar and legends of the Maya civilization*. London: Pantianos Classics.

Spores, R., and Balkansky, A. K. (2013). *The Mixtecs of Oaxaca: Ancient times to the present*. Norman, OK: University of Oklahoma Press.

For an understanding of the impact that smallpox and disease had on Indigenous society in Mexico and throughout the region, see:

Crosby, A. W. (1972). *The Columbian exchange: Biological and cultural consequences of 1492*. Westport, CT: Greenwood.

For a firsthand account of the conquest, see:

Díaz del Castillo, B., and Cohen, J. M. (1963). *The conquest of New Spain*. Middlesex, England: Penguin Books.

To learn about the environmental–societal dialectic of pre-Columbian cultures, see:

Mann, C. C. (2005). *1491: New revelations of the Americas before Columbus*. New York: Knopf.

For an Indigenous account of the conquest of Mexico by the Spaniards, see:

León Portilla, M. (1992). *The broken spears: The Aztec account of the conquest of Mexico.* Boston: Beacon Press.

For an introductory geography textbook on Mexico, see:

Rhoda, R., and Burton T. (2010). *Geo-Mexico: The geography and dynamics of modern Mexico.* Vancouver, British Columbia, Canada: Sombrero Books.

To learn more about the political–economic story of where neoliberalization emerged and how it proliferated across the globe, see:

Harvey, D. (2005). *A brief history of neoliberalism.* New York: Oxford University Press.

## REFERENCES

Global Regulation. (2022). Retrieved May 29, 2022 from *http://global-regulation.com/translation/mexico/1424991/law-general-of-linguistic-rights-of-indigenous-peoples.html.*

Instituto Nacional de Estadística y Geografía (INEGI). (2020). Lengua indígena. Retrieved June 18, 2024, from *http://inegi.org.mx/temas/lengua/#Tabulados.*

Marston, S. A. (2000). The social construction of scale. *Progress in Human Geography,* 24(2), 219–242.

The Newcastle Social Geographies Collective. (2021). *Social geographies: An Introduction.* New York: Rowman and Littlefield.

Panelli, R. (2004). *Social geographies: From difference to action.* London: SAGE.

CHAPTER 2

# Shrines of the Times

## *The Social Geographies of Contested Space in Tepito and the Historic Center of Mexico City*

This chapter analyzes the corollary links among shifting economic policies in Mexico, the proliferation of the informal economy, and the intensification of La Santa Muerte adoration (a Mexican deity not yet recognized by the Catholic Church, but more about her later) in Mexico City's historic center and the Tepito neighborhood. Within this context a description of Mexico City's neoliberal public–private revitalization plan of its historic center, called *El Programa de Rescate* (recovery program), is described, with attention to stakeholders' interest in the production and utilization of urban space.

### Chapter Outline

The chapter begins with a description of Mexico City's location, including a brief history of the city before introducing La Santa Muerte, one of the stars of this story. The Tepito neighborhood and the people who live and work in this fiercely independent part of the city are next introduced. This discussion is followed by an analysis of shifting economic policies in Mexico and their impacts on Mexico City, which correlates with an increase in the informal economy, especially in the Tepito neighborhood, where we learn how Beto, a taco vendor, copes with the realities of neoliberalism. Finally the implementation and successes of the historic center's revitalization plan are discussed. In covering all these topics, this chapter describes one of the many stories of contemporary life in neoliberal Mexico.

## A Brief History of Mexico City

Mexico City, situated in the Valley of Mexico, a highland plateau surrounded by mountains and volcanoes that sits at an elevation of 7,349 feet. The city is geographically located in central south Mexico where it borders the state of Mexico on the west, north, and east and the state of Morelos to the south (see Figures 2.1 and 2.2). This is a big city, in the size of its population as well as in its overall urban area. The city proper, what used to be called the Federal District and now is called Ciudad de Mexico, or CDMX, has a population of 9.2 million, while the conurbation of Mexico City, which spreads over the borders of CDMX into the states of Mexico and Hidalgo and is officially called the Metropolitan Area of the Valley of Mexico, contains 21.8 million people and covers 577 square miles. This may seem like a lot of people per square mile, and while it is densely populated in comparison to other Mexican and Latin American cities, it is not as densely populated as many other global metropolises. For example, in comparison, New York City's population density is over 29.3 thousand people per square mile, whereas Mexico City is nearly less than half at 16,000 people per square mile. The difference in density is due to the built environment; New York City is much

FIGURE 2.1 Map of Mexico, indicating the location of Mexico City.

more vertical than CDMX, which, aside from the financial and tourist center along Reforma Avenue and new residential skyscrapers emerging in the upscale districts in the boroughs of Cuauhtémoc, Cuajimalpa, and Benito Juárez, is a sprawling city of single-, two-, and three-story buildings.

People have been living in cities in the Valley of Mexico for a long time. When the Spanish Conquistador Hernán Cortéz and a thousand infantry first laid eyes on the Aztec capital of Tenochtitlan, the present-day site of CDMX, they witnessed the largest city they had seen in their lifetime: a well-organized metropolis of 140,000 inhabitants. Including the other cities found in the valley, such as Atzacoalco and Texcoco and the smaller towns of Coyoacan and Iztapalapa, nearly 1 million people were living in the Valley of Mexico in 1519. The sight of the city built in the middle of a lake mesmerized Cortés's entourage:

> "When we saw all those cities and villages built on water; and the other great towns on dry land, and that straight and level causeway leading to Mexico [Tenochtitlan, the Aztec city], we were astounded. These great towns and shrines and buildings rising from the water, all made of stone, seemed like an enchanted vision from the tale of Amadis. Indeed, some of our soldiers asked whether it was not all a dream. . . . It was all so wonderful that I do not know how to describe this first glimpse of things never heard or never seen, and never dreamed of before", wrote the chronicler, Bernal Díaz del Castillo. (Díaz del Castillo et al., 2012)

The former Aztec capital that had been built by the Aztecs probably around 1325 was severely damaged in the siege of Tenochtitlan and the fall of the Aztec Empire. It would be rebuilt by the Spaniards and become the colonial political, administrative, and financial center of the Viceroyalty of New Spain and the capital of the new nation of Mexico after gaining independence from Spain.

Mexico City had been considered a Federal District (Mexico D.F. in Spanish) and the capital of the country's 31 states, much like the District of Colombia, or Washington, D.C. In 2016, the Federal District began the transition to become the 32nd federal entity in Mexico, which provides the city a level of autonomy comparable to that of a state with a constitution and legislature. The city's 16 boroughs are now headed by elected mayors with *a jefe de gobierno de Ciudad de Mexico*, or government city boss, with executive power over the city.

The case study for this chapter focuses on the production and utilization of urban space in the historic center and the Tepito neighborhood, which are both located in the borough of Cuauhtémoc (Figure 2.2). It is in Cuauhtémoc where the Aztecs had built their capitol Tenochtitlan, which was the center of power for their empire and which the Spaniards astutely decided to make their own. Indeed, the design of streets and plazas of the historic center that exist today have changed little since the end of the 1600s. It consists of 668 blocks with 1,500 buildings, including all the 17th- and 18th-century colonial palaces remaining from the viceregal era, 196 civic monuments, 67 religious monuments, 53 museums, 68 plazas, 19 cloisters, 28 fountains, and 12 public murals.

During the Colonial period the historic center housed the Spanish dignitaries, administrators, and the *criollo* elite merchants. Indigenous peoples and Africans would have lived in districts outside the center. After gaining independence, this pattern continued, with elites residing exclusively in the *centro* where administrative offices and sought-after amenities could be found.

It would not be until the last decade of the 20th century with the introduction of street cars and a concerted urbanization effort by then President Porfirio Diaz that an elite corridor, Reforma Avenue, would be built west of the *centro* replete with the upper-class neighborhoods of La Condesa and Roma. The pattern of building upper-class neighborhoods west of the *centro* continued after the Mexican Revolution with the

FIGURE 2.2 Map of greater Mexico City, highlighting the Cuauhtémoc borough and the location of the historic district.

urban development of the Polanco neighborhood, followed by Chapultepec, adjacent to the city's massive Chapultepec Park. With the introduction of automobiles and the construction of autocentric infrastructure, the wealthy continue to live farther west in the now elite neighborhoods of Santa Fe and Satélite.

With each wave of urban development toward the west, the elites began to gradually abandon the centro. Its palaces subdivided into tenements became affordable housing for the millions of migrants who came to México City in the 20th century. The *centro* became a staging ground, or starting point, for migrants to become settled in the city before occupying land and constructing a home in one of the burgeoning shanties toward the east and south of the city, such as Netzahualcoyotl.

An 8.0 earthquake struck Mexico City on September 19, 1985, at 7:17 in the morning. Mexico City is built on the bed of the historic Lake Texcoco whose silt and volcanic clay sediments amplify shaking from the quake, signifying damage to structures worsened by soil liquefaction. The borough of Cuauhtémoc, which includes the historic center and Tepito, suffered enormous damage with 258 buildings completely destroyed, 143 partially collapsed, and 2,000 structures seriously damaged; 10,000 people were killed.

The Ministry of Foreign affairs refused to accept aid from the United States and then-President Miguel de la Madrid decided against putting a halt on foreign debt payments so that funds could be used to help with the recovery effort. Instead, the government initially responded with de la Madrid declaring a period of mourning for 3 days, then deployed the military to prevent looting after a curfew was imposed. Reports cite that, in the crucial first days for recovery immediately following the earthquake, the military assisted factory owners in retrieving their machinery rather than removing the bodies of dead factory workers. Multiple segments of Mexican society viewed the PRI's response as both authoritarian and incompetent.

At the time of the 1985 earthquake Mexico was in its fourth year of a foreign debt crisis and a contracting economy that caused political problems for the PRI. Mexico's one-party rule throughout the post-revolutionary 20th century was tolerated because the country experienced four decades of economic expansion. When this ended with the debt crisis beginning in 1981, the PRI machine's power began to wane. The debt crisis, coupled with the PRI's poor mismanagement after the 1985 earthquake, provided an opportunity for grassroots organizations to step in, organize, and help.

The people from Tepito quickly demonstrated their ability and willingness to rebuild by themselves. They began to tear down unsafe buildings by hand, carefully preserving the building materials for later reconstruction. They rallied together and demanded that the government provide them the funds so they could repair their damaged homes themselves. The Tepito residents collaborated with representatives from 150 other communities and autonomous organizations to form a self-help network to facilitate the circulation of information, talents, and government resources. They successfully fought for the expropriation of damaged buildings. The crisis of the earthquake provided an opportunity for Tepito residents to become more autonomous and further contributed to its tightly knit identity (Matas Rosas, 2010).

The palaces that were turned into tenements that had emerged in the *centro* were badly damaged by the earthquake. The population decreased, many of the palaces remained unrepaired and vacant, and the historic center's streets became open-air markets spilling over from the informal commerce of street vending in Tepito. In many ways the deterioration of the historic center in the years following the earthquake paved the way for the revitalization program that would be implemented in 2002 and continues until today (Smith, 1987).

Against this backdrop of the postearthquake crisis of the second half of the 1980s, the historic center and Tepito witnessed an increase in street vending and La Santa Muerte adoration. The next section discusses La Santa Muerte.

## What Is La Santa Muerte?

> "*La Sante Muerte* is part of folkloric creativity [*la creatividad popular*]. You can see this creativity in how they adorn images [of *La Santa Muerte* statues], how they dress her as characters from fairy tales, in wedding dresses, in *chinas poblana* (A colorful outfit first worn by the women of Puebla comprising a brightly embroidered white blouse worn with a red and green skirt with sequins) . . . people make images of *La Santa Muerte* from paper, from herbs, from cardboard, from lead, from seeds, from soap. . . . I mean, like, there are no limits to this creativity." (Gerardo Cabrera, Colaborador de la revista Devocion a la Santa Muerte, quoted in Adeath and Kristensen, 2007, p. 16)

There are many forms of cultural and religious syncretism in Mexico from the Day of the Dead—a celebration on November 1 and 2 that honors deceased friends and relatives—to the worship of the patron saint of Mexico—the Virgen de Guadalupe. So too is La Santa Muerte a product of religious and cultural syncretism. The pre-Colombian Mexican predecessors of the Santa Muerte are the god and goddess *Mictlantecuhtli* and *Mictecacihuatl*, the deities of death and darkness in the Aztec "religion of the dead." *Mictecacihuatl* presides over Mictlan, the underworld. The followers of these deities provided *ofrendas*, or offerings, for the gods to arrange favors for them—just as the Santa Muerte is currently worshiped. Manuel Valdéz, a spiritual counselor who works at Sonora Market in the Merced Balbuena neighborhood of the historic center explains, "Our ancestors piously worshipped (*creían mucho*) *Mictlantecuhtli*, who was the god of death. Since we are devout to La Santa Muerte, we interpret her to also be a deity" (M. Valdez, personal communication, July, 2017).

Likewise, in Europe, the pre-Christian worship of the earth goddesses in the multitheistic religion of Mithraism practiced by Roman soldiers shifted—through a form of pagan and Christian syncretism—to the worship of the Virgin Mary and multiple other virgins found across the Catholic countries of Europe. The worship of virgins is referred to as *Maríanismo*. In West-African religions there are many goddesses, called orishas, who represent the earth and sea—such as Oshun—associated with water and purity. The belief systems of Africans who were forced to come to the Americas to labor as slaves included these West African deities that would syncretize with Christianity to become Santeria (Wolf, 1958). The Santa Muerte belief system borrows from each of these religious traditions. As expressed by Doña Queta, the keeper of the altar at number 12 Alfarería Street in Tepito, Santa Muerte is a syncretized deity as "Here we do not celebrate mass [*no hay misas*], there are no devil worshipers, nor are there Christians. There are no *misas* because this is an altar that does not have the recognition of the Catholic Church, so it is a place [site] where people come to pray, they make request [*pedir que la hacen un paro*, the Tepiteño way to ask for a favor] to and provide offerings to the Santa Muerte. You won't find any devil worshipers here because we are not part of a satanic sect or are tied to any form of devil worshipers" (D. Queta, personal communication). The Santa Muerte is a combination of these pre-Colombian beliefs with southern European *Maríanismo* and a dash of Caribbean *Santeria*.

Life-size figures of the Santa Muerte stand in front of stores, bars, and restaurants lining the streets of Tepito (see Figure 2.3). Doña Queta and her family marked Tepito as the territory of La Santa Muerte come "to life" by constructing the public altar created in her honor at 12

FIGURE 2.3 La Santa Muerte statue in front of a bookstore and ice cream shop. Photo by David M. Walker.

Alfarería Street in Tepito. Additionally, many of the buildings are decorated with Santa Muerte murals and graffiti. Her image is part of the urban fabric of Tepito. The construction of public altars has spread from Tepito to the rest of the neighborhoods surrounding the historic center and now can be found across Mexico, Central and South America, and the Philippines.

Devotees bring La Santa Muerte into the private sphere of the home as well, where people construct domestic altars for her. Don Ray describes how at the altar he has built for her at home, he offers "water, honey, sugar, salt, and whatever is close at hand—sweets, an apple, some wine . . . these things are what nourishes her and provides her abundance," demonstrating the frugal nature of the offering, but he believes these things bring her "abundance" (D. Ray, personal communication).

La Santa Muerte is also venerated in a physical way. Women and men tattoo her image across their torsos and forearms or dangle small amulets of the figure from necklaces that tangle with crucifixes or small talismans of the Virgen de Guadalupe. I also saw a French woman in Port Lucate in the South of France near the Spanish border with a tattoo of

La Santa Muerte, demonstrating her international appeal. The number of worshipers has grown in Mexico City and beyond.

The Mexican poet and philosopher, Homero Aridjis, reinforces that the Santa Muerte, which he insists is a contemporary form of religious syncretism, "is rapidly growing, the [adoration]of the Santa Muerte is intertwined with the Day of the Dead, it has at its roots pre-Hispanic beliefs as well as colonial traditions. . . . [Santa Muerte worship] lends itself to contemporary urban Mexico" (Aridjis, 2017, p. 67). Tepito residents feel so strongly about La Santa Muerte, they choose to adorn their body with tattoos of her image as the young man has in the in the image in Figure 2.4.

These socially produced public and private altars, murals, and statues to La Santa Muerte are reproducing the landscapes of the neighborhoods where she is adored by and for her devotees. They function as a mechanism to perform the identities of and build solidarity among her followers, while they also reveal how the working poor appropriate and reproduce urban space in these locales.

Virgen de Guadalupe, the patron saint of Mexico recognized by the Catholic Church, is also a product of religious syncretism that was

FIGURE 2.4 Man with a tattoo of La Santa Muerte on his back. Photo by David M. Walker.

originally banned by the Vatican. Legend or history holds that an Indigenous peasant, Juan Diego, first found the shroud with the image of the virgin in 1531. Mexicans adored her for over 200 years until she finally received pontifical approbation in 1754, became sanctified, and was named Our Lady Patron of what was then New Spain.

If La Santa Muerte's popularity increases and her adoration continues to diffuse across Mexico, Latin America, the Philippines, the United States, and beyond while the globe continues to urbanize, will the adoration of the Santa Muerte—this cult that lends itself to contemporary urban life—one day be accepted across greater society and recognized officially by the Vatican with pontifical approbation? The historic pragmatism of the Catholic Church and the Vatican indicates that she will. The next section introduces the Tepito neighborhood and the people who live, work, and worship in this fiercely autonomous area.

## Santa Muerte Worshipers in the Tepito Neighborhood

The historic center's Tepito neighborhood comprises 72 blocks occupied by 120,000 inhabitants, an estimated 15,000 more who come to the neighborhood to sell food and wares, while thousands more shop there daily. It is home to the main shrine dedicated to the adoration of the Santa Muerte.

The neighborhood is one of the centers of the informal economy in Mexico City. The informal economy consists of the diversified set of economic activities, enterprises, jobs, and workers that are not regulated or protected by the state. The concept originally applied to self-employment in small, unregistered enterprises. It has been expanded to include wage employment in unprotected jobs. Informal workers are trying to earn an honest living against great odds. The neighborhood's primary economic activity consists of informal street vending, clandestine manufacturing, and the selling of *fayuca*, or smuggled products.

The Tepito neighborhood traces its history to the preconquest years. The etymologies of the name come from the Nahuatl *teocali-tepiton*, meaning small temple or chapel. The village remained politically outside of Tenochtitlan–Tlatelolco (the Aztec names given to the two independent cities that would become Mexico City). The area became a staging site for people bringing their goods to the larger, Aztec-sanctioned markets. Carriers would stop in Tepito, buy food, and rest; some stayed overnight if they were traveling long distances. These carriers hauled goods that the *Tepiteños* wanted to buy. The carriers sold goods in Tepito and thus did so without paying tribute or taxes to the Aztec tax collectors in the sanctioned markets: Tepito's association with informal markets was born.

At the beginning of the Mexican Miracle, the government of the city "froze" the rents of low-cost housing. The people of Tepito did not perceive this measure as temporary, particularly since they struggled for it. Consequently, this special arrangement exists even today, despite countless attempts by lawyers, politicians, and developers to eliminate it. Those who live in Tepito enjoy an economic privilege compelling them to remain. Their rents are among the cheapest in the city.

Tepiteños conquer spaces with ingenuity. They get a second floor by building one in the interior of their houses. The in-situ accretion architecture is representative of the informality and organic growth of Mexican cities during the neoliberal age. Many houses serve as workshops during the day and as homes at night. Houses and entire buildings have been transformed into studios for selling pirated music, movies, books, and clothing. Patios are common spaces with multiple purposes.

Tepiteños have transformed the streets into places for workshops, trade, and recreational activities and have reproduced Tepito into a creative and recreative space. The trade in used clothes flourishes next to that of new clothes sewn in Tepito. Shoe repairmen prosper next to workshops that produce new shoes. They have remade, remodeled, and transformed with ingenuity a thousand mechanical and electrical gadgets thrown out by their wealthy owners or smuggled across the border from the United States. The objects reformulated by Tepiteños have become famous for their value and quality, which is well known throughout the metropolitan area.

An important institution in Tepito is the Centro de Estudios Tepiteños, or the Center for Tepito Studies (CET), located on Jesus Carranza Street. Alfonso Hernandez is a deprofessionalized academic, with degrees from UNAM in history and philosophy. He is a self-proclaimed specialist and *cronista* of Tepito who founded the newspapers *El Ñero* and *Tepito y Anexas*, the magazine *Tepito Sí*, and the CET. The CET is a think tank and public relations office that promotes the folklore and traditions of Tepito and works to preserve the livelihoods of street vendors. Hernandez was born and raised in the heart of the neighborhood and, as he says, with an "*albur, para que la cuña apriete, tiene que ser del mismo palo*"—or for the wedge to stick, it has to be from the same plank (i.e., it takes a thief to catch a thief), demonstrating that he is well suited for his position as director.

*Albures* are a significant component of the Tepito identity and sense of place. The *albur* is a linguistic mode of the Spanish language, popularly yet marginally used throughout Mexico, that serves to construct alternative identities, understandings, and a sense of community that emerged in Tepito (Anaya, 2014). *Albures*, or alternative utterances, function to unite street vendors and Santa Muerte worshipers. The director of the CET explained how the florid language of *albures* functions to prevent Tepiteños from becoming displaced and possibly eradicated by the system. The director elaborates how "the language used in Tepito forms part of the barrio's identity, creating a cohesive-interconnected community of resistance to the development strategies of the government—and allows the residents of Tepito to also resist modernity [i.e., urban revitalization]" (personal communication, 2017).

The utilization of language to thwart authority or to create confusion, difference, or multiplicity within the postcolonial urban spaces of Tepito has historic antecedents. According to Victor Hugo Rocha, a self-proclaimed distinguished scholar at the CET, "when the necessity arose for the Mexicas [the original inhabitants of Tenochtitlan] to speak in front of the Spanish conquistadors they chose to develop verbal 'double entendre' codes" (V. H. Rocha, personal communication). In contemporary urban México, the putting into practice of *albures* allows for the special production of nodes of resistance. The vernacular form of language used in *albures* is doing more than just "playing with language." In regard to language and identity, the construction of *Mexicanidad*, or Mexicanness, the Mexican philosopher Octavio Paz states, "in this occasion it is demonstrated that the *albur* is not only an ingenious, whimsical play on words or verbal combat threaded together with obscene allusions and double entendres, in which the winner is he [or she] that is able to contest and respond. If not, he [or she] is possessed and violated by the other. It is the formation of lower [México Profundo] class *Mexicanidad*" (Paz, 1961, p. 35).

Like Mestizos—who are a product of *La Malinche* and the conquistador—Tepiteños are using the deterritorialized languages of both the Spanish invaders and the Nahuatl spoken by the Mexicas or Aztecs. *Albures* are a dialogical form of revelry that exists to challenge imposition, to use language for appropriation, and to counterattack the rest of the world. The language of the Tepiteños is infused with place-based identity politics.

The residents of Tepito have continuously reproduced their neighborhood identity in the face of mainstream Mestizo Mexicanness. The Tepiteños comprise one of the many groups that encompass what the Mexican philosopher Guillermo Bonfil Batalla calls the "México Profundo"—or the "deep," "profound," or "true México." Alfonso Hernandez insists that Tepiteños through their construction of identity around the *albur*, the informal economy, La Santa Muerte, and vernacular food practices, such as *migas* and *tacos de vísceras*—dishes made from stale bread and other ingredients like intestines, or tripe—they are producing the spaces of production and consumption in Tepito for Tepiteños.

Tepito teems with activity. People march through the streets in search of vendor stalls in the open-air markets (Figure 2.5), eat at the many taco carts and food stalls, and gather around *tepache* (a fermented beverage made from the peel and the rind of pineapples) and beer vendors to quench their thirst and discuss business, politics, and the predicaments of friends and family (Figure 2.6).

**FIGURE 2.5** Busy shopping street in Tepito. Photo by David M. Walker.

FIGURE 2.6 Tepache/soda pop/beer vendor in Tepito. Photo by David M. Walker.

Many are also on route to pay tribute to the Santa Muerte at her main altar at Number 12 Alfarería Street (Figures 2.7 and 2.8). Here, single mothers, shop owners, taco vendors, garbage pickers, drug dealers, and even police—people from all walks of life—arrive to offer the Santa Muerte cigarettes, money, tequila or mezcal, her favorite foods, drugs, and photographs or to sing songs and recite poetry to her in exchange for help with love, work, and health issues, but also to seek revenge upon a foe or remove a friend or family member from harm's way. At the shrine they will make a pact with Santa Muerte. They do not pray to her. They negotiate: *Haz mi paro*—a *Tepiteño albur,* a way of saying, "come on, do me the favor, I'll get you back." Tepiteños care for and worship at the Santa Muerte shrine daily, while believers travel from across town or the nation to pay their respects, ask favors, and form pacts with her.

Santa Muerte worship is economically viable for the working-poor street vendors. For example, an *ambulante* (street vendor) described how "people nowadays are losing their faith [in the Catholic Church] . . . they no longer trust the Catholic priests who charge for everything. And

FIGURE 2.7 Inside the shrine at 12 Alfarería Street, Tepito (notice the woman reflected in the window looking at *La Flaca*). Photo by David M. Walker.

FIGURE 2.8 Street view of the shrine. Photo by David M. Walker.

don't even talk about the [Pentecostal] Christians, who up front don't charge for anything but afterwards pass the plate demanding large sums of money. That is why I lean more toward La Santa Muerte, who listens to me without paying her money" personal communication). Doña Queta also indicates that economically challenged residents of Tepito have turned to Santa Muerte because " . . . in *Santeria* practice you spend a lot of money. For example, if I tell a *Santero* [a *Santeria* priest] I want you to go to my house and cure [purify] my *santitas* [little saints, the small statues of saints recognized by the Catholic Church], he's going to tell me that for this saint we have to sacrifice a goat, for this one a turkey, and for that one two black chickens. Additionally, on Monday I have to pray to one saint for an hour in the morning and to another for an hour in the afternoon . . . and on Wednesday its going to be another saint . . . and that's how the days continue. I don't have the time. With *La Flaquita* [a nickname for La Santa Muerte], it's different" (D. Queta, personal communication, 2017). Here Queta reveals the compatibility of *Santeria* and Catholicism and reveals her monetary and practical preference for La Santa Muerte over *Santeria*!

Most altar keepers dedicated to La Santa Muerte are women, which within a patriarchal society reveals how the deity attracts and is venerated by the marginalized. This practice creates an interesting power dynamic, wherein the public gatekeepers of Santa Muerte worship are not dominated by male leaders, "but by matrons [*matronas* in Spanish, which comes from the Latin *mentors*], where these *madrezotas* (a Tepito *albur* for "big obstacle" or "huge thing," literally big bad-ass mothers) who *direct* the cult yet are not responsible for the cult's ideology. Doña Queta, for example, leaves [that responsibility] to the people . . . declaring, don't ask me, ask the people [i.e., the worshipers]" (Journalist, quoted in Adeath and Kristensen, 2007, p. 41). Doña Queta and her family are praised for having an altar that "imposes respect." Her social skills, knowledge of the neighborhood, and fluency in *albures*, has helped her earn respect for devotees and people with no faith in La Santa Muerte.

Tepiteños and people across Mexico turn to Santa Muerte worship as the realities of daily life under neoliberal governance have become harsh, unpredictable, and more difficult. In Tepito, residents have expressed how they feel both abandoned by the state and by municipal policies that seek to thwart their livelihood strategies as vendors in the informal markets. This worship supplements residents praying to and asking favors from the Virgen de Guadalupe. Worshipers seek a deity who is outside of the state-sanctioned system and not affiliated with the Catholic Church. This adoration, in combination with *albur* slang, the informal economy, and vernacular food practices function as an apparatus in the configuration of a counterculture identity that resists linear development policies (see Figure 2.9 for an example).

The increase in La Santa Muerte worshipers from the 1980s onward has occurred in tandem with the escalation of urbanization, underemployment and poverty, and the weakening of the centralized state in Mexico City and across the nation. The next section contextualizes why and how these social geographies emerged. It requires an explanation of the shift from populism to neoliberal governance to understand the relationship among street vending, the informal economy, Santa Muerte worship, neoliberalism, and the negotiation and resistance to neoliberal practices within the socially produced spaces of Tepito and the historic center of Mexico City.

**FIGURE 2.9** La Santa Muerte at the door of a bar in the historic district. Photo by David M. Walker.

## Rapid Urbanization and Economic Shifts from ISI to Neoliberalism

Mexico rapidly industrialized and urbanized from the 1950s through the 1970s in tandem with the implementation of import substitution industrialization (ISI), consisting of restrictive import tariffs, especially on manufactured goods. ISI policy worked to protect Mexico's domestic manufacturing. During the experiment with ISI, factories in Mexico churned out electronics, and domestic goods, such as shoes and clothing, and even automobiles—the famous *Vochos*, or Volkswagen Beetles, manufactured in Puebla under agreement with the German Volkswagen Group that the cars were to be sold solely in Mexico. All these goods were produced in Mexico, by Mexicans, for the domestic market. The decades of the 1960s and 1970s witnessed Mexico's economy increase 7 percent annually, with manufacturing rates at full capacity employing millions of workers. These were the golden decades in Mexico's development, the *Milagro Mexicano,* or the Mexican Miracle. ISI policy produced the industrial urban landscapes of post-1950s Mexico City.

Urban consumers purchased the domestically manufactured goods. To lower transportation costs the state gave incentives for the construction of manufacturing centers in and around cities. Because of urban primacy in Mexico, and due to the important political, cultural, and economic roll Mexico City has played in the creation of the nation, Mexico City became the primary site of industrialization (although Guadalajara and Monterrey also became industrialized and urbanized during ISI). As the factories grew, more jobs were created, and rural-to-urban migration produced the emerging urban landscapes of Mexico City. Mexico City's population more than doubled between 1960 and 1980 from 6.3 to 13.5 million residents. The urbanization processes during this period witnessed the burgeoning upper-middle class moving to neighborhoods and suburbs west of the historic center. Informal urbanization processes were produced in the historic center and its surrounding neighborhoods by rural migrants coming to and using the city as a starting point for the eventual move to the self-help housing east of Mexico City. On the dry lakebed of Texcoco the large irregular settlement Netzahualcoyotl emerged, coupled with other irregular settlements in the south, especially among the Ajusco foothills.

## The Implosion of ISI

With the discovery of vast new oil beds in the Gulf of Mexico in the late 1960s, oil exploration, drilling, and exports began to shadow the domestic manufacturing sector of the economy. By the mid-1970s Mexico's economic outlook appeared bright. Indeed, the economy grew 8 percent annually from 1978 to 1982. A strong domestic manufacturing and consumer sector and vast oil reserves to be harvested and exported to its northern neighbor paved the way to national prosperity.

In the decade of the 1970s oil prices were high, and international banks flush with money were eager to recycle their cash by lending it to developing countries. Although Mexico had large oil reserves, it needed capital and new technologies to exploit them. The country borrowed heavily to do so, believing that the high oil prices would remain, and the debts could easily be paid off.

However, by 1982 the Reagan administration had busted up OPEC (the Organization of Petroleum Exporting Countries), oil prices dropped, and interest rates increased in the United States and Europe. Debt payments also increased, making it harder for Mexico to pay back what it owed international lenders. Mexico's main money earner, exported oil, decreased in value while payments to international lenders to finance its debts—money it had borrowed to grow the oil industry—began to increase exponentially. By 1982 Mexico could no longer continue to pay off its external debt, and it became the first country to default on its loans. The IMF and the World Bank stepped in to reorganize the economy.

The era of ISI abruptly ended, and the Mexican Miracle faded with the opening of the economy through the adaptation of neoliberal economic policies. *Neoliberalization* consists of a set of discourses and processes based on the belief that the market is the ideal mechanism for the organization and regulation of social, economic, and political life (Walker, 2008). Grounded on the laissez-faire theories of classic political economy, the emergence of neoliberal policies is associated with the structural adjustment programs (which are "free market" economic policy reforms imposed on developing countries) and the adoption of austerity measures (which are attempts to significantly curtail government spending) of the IMF and the World Bank. These policies and programs include the elimination of subsidies and tariffs; the rationalization of agriculture and the deregulation and privatization of industry; making transparent market exchanges; allowing currencies to float freely; the dismantlement of social programs devoted to education, health, and housing; the downsizing of public employment; and the implementation of strict monetary policies to control inflation. The shift under these policies to neoliberal governance was not compatible with ISI.

**VIGNETTE: Migration, Emigration, and Neoliberalism**

The neoliberal policies and the undermining of the agricultural sector that occurred at this time spurred rural-to-urban migration to Mexico City and to emigration to the United States. From 1980 to 2000 Mexico City grew from roughly 12 million to 21 million people, with nearly a 4.5 percent growth rate per year throughout the 1980s, fueled by peasants leaving rural areas to look for jobs in the city. It is hard to measure the number of *Mexicanos* (those who are Mexican born) living in the United States. The Pew Research Center estimates there are more than 11 million Mexican-born immigrants (both documented and undocumented). In 1980 there were around 1 million. There is a transparent correlation between the implementation of neoliberal policies, a decrease in the agricultural sector, increases in underemployment in Mexico, migration to the United States, and population growth in Mexico City.

Because the state loses control under these policies, they are referred to as elements of *roll-back neoliberalization*, a term coined by geographers Peck and Tickell (2002). Mexico began to adopt roll-back neoliberal policies in 1982. This initial stage of roll-back neoliberalism (roughly 1982–2000), when the state retracts and the social safety net diminishes, correlated with an increase in the popularity of Santa Muerte worship among the urban working poor in the informal sector and the neoliberalization of urban space through an increase in informal economic activities. It is within this context of roll-back neoliberalization that the urban working poor have implemented livelihood strategies of street vending and Santa Muerte adoration to appropriate urban space in the historic center and Tepito.

As Peck and Tickell (2002) note, however, from the early 2000s on another distinctive phase called *roll-out neoliberalization* began to occur. Roll-out neoliberalization consists of aggressive intervention by governments around issues such as crime, policing, urban policies, and surveillance, with the purpose of disciplining, containing, and relocating those marginalized or disposed by the neoliberalization of the 1980s and

1990s. Roll-out neoliberalization often involves state collaboration with the private sector to implement urban policies and to renovate and police perceived derelict areas—such as the historic center and surrounding neighborhoods—that were thrown into disarray during the earlier period of roll-back neoliberalization. Mexico City's urban revitalization program provides an example of a roll-out neoliberal urban policy. The implementation of the *programa de rescate*—the private–public, roll-back neoliberal revitalization plan—in the historic center and its surrounding neighborhoods aims to discipline street vendors and Santa Muerte worshipers. In these locales the contestation of urban space occurs (Figure 2.10).

U.S.-trained Mexican economists and politicians, called technocrats because they advocated for policies learned at the University of Chicago from the trained technical economists, fully embraced neoliberal doctrine. The adoption of these policies has correlated with a decrease in the standards of living for millions of Mexican people. Indeed, wages for Mexican workers have lost 82 percent of their purchasing power since 1982, the year when neoliberal reforms were first imposed on Mexico. The IMF structural adjustment programs, in conjunction with NAFTA in 1994, undermined Mexico's food security by mandating the privatization or dissolution of state-regulated grain reserves and price-support programs, sweeping reductions in farmer credit and subsidies, the deregulation of commodities markets, and the elimination of tariff and quota protections on imported agricultural products.

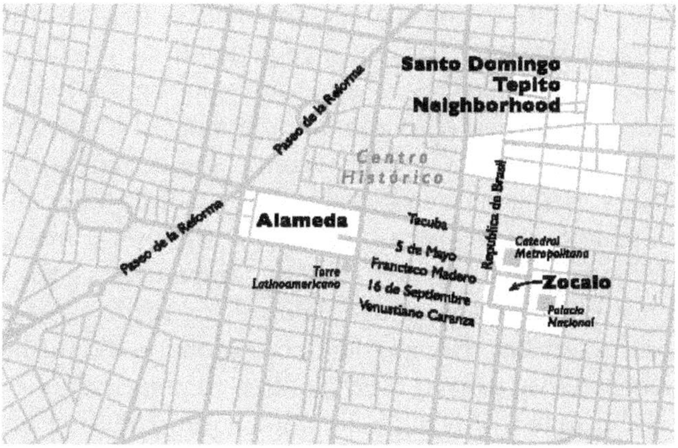


FIGURE 2.10 Map delineating the important sites across the historic center. Map by David M. Walker.

The result of neoliberal policies in Mexico has correlated with a rise in food, clothing, and commodity prices and a shift in manufacturing jobs located in the center of the country, which produced goods for Mexican consumption, to export-oriented manufacturing zones, called *maquiladoras*, located along the U.S.–Mexico border (Mexico City lost over 385,000 industrial jobs between 1982 and 1993). The adaptation of neoliberal policies, the near elimination of federal and municipal poverty alleviation programs, and shifts in manufacturing from the center to the north correlated with a rise in unemployment, underemployment, and poverty in Mexico City. The economy retracted throughout the 1980s, while the Mexican peso would be devalued by 500 percent, giving this period the designation of the Lost Decade.

As the neoliberal experiment progressed and formal employment decreased, the Mexican people became fed up with the inability of the state and the market to supply goods, services, and employment to its citizens. The next section discusses how the informal economy functioned as an escape valve for the unemployed and underemployed, as people turned to the informal economy as a survival tactic.

## The Production of the Informal Economy

A component of roll-back neoliberalization is the creation of a hyper-efficient government sector. This requires that government be downsized to make labor pliable to the market. In Mexico, government agencies employed large segments of the educated populations, providing incomes to the burgeoning middle-class communities during the golden decades of the Mexican Miracle. With the elimination of high-paying government jobs, a huge swath of middle-class-income earners was wiped out from Mexico City's economic picture. The eradication of high-paying jobs in the manufacturing sector, coupled with fewer government payroll jobs, reverberated across the tertiary components of the economy: Service-sector employees were also laid off, as there was no longer a demand for the services they provided; construction diminished; people could no longer afford to send their children to university.

How did people in Mexico City cope with increasing poverty and the rollback of government-sponsored programs? They looked toward and literally created the informal economy for their livelihood strategies, which also involved the reproduction and utilization of urban space.

Residents from all walks of life resorted to street vending to cope with the underemployment and unemployment associated with neoliberal economic reforms initiated in the 1980s that continue to the present. Opting for work in the informal economy boosted economic activity in the street and changed the use of urban space across the historic center. Although vendors always existed in Mexico City, during this time Tepito and the historic center became associated in the minds of the city's residents as the place to go to buy pirated goods at a fraction of the price they cost in stores.

In Mexico City, and especially in Tepito, the informal sector of the economy is pervasive in that it covers the clandestine manufacturing of goods, consumption, where to find a job, where to meet people, health care, recycling, and housing, resulting in the reproduction of urban space by the working poor. Not including drug trafficking and human smuggling, it consists of the production of everything from shoes and clothing in clandestine factories, the preparation and selling of food, hawking phone cards, constructing houses, making furniture, smuggling and selling household goods imported clandestinely from China, to the buying and selling of stolen goods smuggled into the country (e.g., cars), and especially the pirating of movies, music, software programs, and even school textbooks. Why buy your university's business-management textbook at full price when you can get it on the street in Tepito at a fraction of the cost?

The Tepito neighborhood is home to the largest book, music, movie, and software pirating infrastructure in Latin America and only second in the world to China. Although it is hard to obtain concrete numbers for the informal economy in Mexico City, Mexican architecture professor José Manuel Castillo Olea claims that 60 percent of housing construction occurs informally; over 60 percent of jobs found in the city are in the informal economy; 25 percent of the 105,000 taxis are unofficial; 25,000 street vendors operate in the historic center alone (this number has increased by 40 percent in the last 15 years); and he estimates that 65 percent of music sold in Mexico is pirated music. According to the Organization for Economic Cooperation and Development (OECD), the informal economy accounts for 57 percent of urban jobs and as much as $600 billion a year—while exports, excluding goods produced in the *maquiladoras*, amount to $400 billion—and make up of 26 percent of the nation's GDP (as compared to the United States where the informal economy is 0.8 percent of GDP).

These goods are sold on the streets by *ambulantes* (street vendors). Sidewalks and entire streets make up spaces of consumption, which literally recreate the utilization of urban space in Tepito. The informal economy and street vending are not new to Mexico or Tepito. The practice of street vendors selling wares on the streets of towns and cities can be traced to the advent of the *tianguis*, which is a Nahuatl word to describe the spontaneous informal markets of the Mexican Highlands, including Mexico City. The *tianguis* have always been "ambulatory" in the sense that these impromptu marketplaces do not always set up in the same location. The bazaars shift and flow according to where there exists a critical mass of people in the towns and cities who demand to purchase the street vendor's goods.

It is important to note that the informal economy positively impacts working poor consumers. Suppressed wages and underemployment mean that even the folks in Mexico City who work in the formal sectors of the economy—such as at the hundreds of *Oxxos* (the equivalent of dollar stores) that have sprung up across the city—prefer to purchase shoes and clothing, household goods (e.g., refrigerators and toasters), and entertainment (TVs, DVDs, books, and magazines) in Tepito rather than at chain stores like Walmex (Walmart is the largest private employer in Mexico).

For example, in Mexican K–12 schools, parents are responsible for purchasing school uniforms, textbooks, and classroom supplies (as a result of the neoliberalization of the education system, which is occurring, too, in the United States). At the beginning of the school year, it is a common sight to see families making their back-to-school purchases in the streets of Tepito. Testimonies from mothers and fathers reveal they couldn't afford to send their children to school if they had to buy what they needed in the corporate department stores. Working-poor single mothers with children don't view the street vendors as breaking the laws by selling pirated goods in the streets of Tepito; they claim they are a godsend! The following section describes how an educated person who once aspired to the middle class had to cope with the neoliberal shift and its impacts on his life.

## Coping with Neoliberalism

The story of Beto provides an example of coping with neoliberalism in Mexico City. Beto earned a bachelor of arts (BA) from UNAM in public administration in 1978. With degree in hand, he quickly found employment as an administrator in what was then Mexico's environmental protection agency, SEMARNAT's (*Secretaría de Medio Ambiente y Recursos Naturales*) offices in Mexico City. He stayed with the same organization through the late 1980s, becoming firmly entrenched in its bureaucracy. The position afforded Beto—who had a beautiful wife with great kids—most of the trappings of what was then a solid middle-class Mexico City lifestyle: a car, a small southern California suburban ranch-style house in the Satélite neighborhood, good benefits, health care, paid vacation, a time share and trips to Acapulco, eating out in restaurants, and Sunday excursions to the countryside. He had improved his position in the world beyond what his parents, migrants who came to the city from Guanajuato in the 1950s, had achieved working in factories. He was educated, had lots of purchasing power, and expected that his children would accomplish even more than he had with his good fortune and perhaps become doctors or lawyers.

Life changed drastically for Beto and his family during the period of roll-back neoliberalism in the 1990s. When I met Beto in 1996 it had already been 7 years since he had full and formal employment using his degree in public administration. He was working driving an unlicensed Volkswagen Beetle (Vocho) taxi at night (a dangerous occupation) and selling *tortas de carne asada* (delicious Mexican grilled steak sandwiches) with his family during the day in the historic center. He had sold his house. It was too far to commute to his current informal employment possibilities. The family car was now his taxi. He moved his family to a working-class neighborhood in the southeast of the city called Iztapalapa. Beto and his family no longer vacationed or took excursions out of the city; they mostly worked. They relied on the free health care from the Mexican government-run hospitals, where he claimed that people mostly go to just twice in their lives, "to be born and to die."

By 2005 Beto had been robbed three times driving his clandestine cab and was hospitalized once when he was stuck in the back of the head, robbed, and left unconscious and bleeding in the taxi until some folks happened upon him and luckily and kindly took him to the hospital rather than leaving him to die in the street and making off with his car, as he explained. At the time of this writing, Beto is 63 years old. He no longer has his clandestine taxi. City officials have come down hard on regulating transportation and for reasons of safety for the drivers and clients and to deter emissions. More and more cabs are hybrids, and the city is working hard to get cars and buses off the roads and people to use the megabuses and the ever-expanding metro. Beto's children didn't go to the university. His son, Beto Junior, still sells tortas. His daughter, Marcela, married a cabinet maker (also part of the informal economy) and is doing well, living in an irregular settlement called Nezahualcoyotl, the world's largest megaslum built on the dry lakebed of Texcoco that became its own municipality in 2006 and is prone to flooding. Beto's experience of being foisted from his middle-class dream is one story among millions across Mexico City. Little did he know he was living out the end of the Mexican Miracle, during a period of multiscalar socioeconomic shifts from the global to the regional and national that directly impacted his life. His negotiations with neoliberalism are emblematic of the Mexican experience during the era of the Shrines of the Times. Beto's son wears an amulet of the Santa Muerte and thanks her for keeping him, his father, and his family out of harm's way in a Mexico wrought with challenges and uncertainty.

## Saving the Historic Center? But from Whom and for Whom?: El Programa de Rescate del Centro Histórico

In this section we turn to the private–public municipal *Plan de Rescate* (rescue plan) to revitalize Mexico City's historic center. This plan provides an example of roll-out neoliberalization of urban space (see Walker, 2008, in "Suggested Resources"). The intensification of La Santa Muerte adoration and street vending emerged with the rollback of the state with the shift from populist to neoliberal governance that began in 1982. The current renovation plan aims to roll out and discipline the people who have developed livelihood strategies and worshipping practices that helped them to cope with life in neoliberal Mexico, which has correlated with the urban working poor to use street vending and the Santa Muerte adoration to appropriate and recreate urban space in Tepito.

Roll-out neoliberalization in Mexico City involves state collaboration with the private sector to implement urban policies and to renovate and police derelict areas that were thrown into disarray during the earlier period of roll-back neoliberalization. The urban policy seeks to renovate buildings, update infrastructure, promote the proliferation of public institutions (like museums and the university), job training and professionalization of low-income populations, and the removal of informal-sector laborers from the public spaces of the historic center. The goal of the rescue plan is to pave the way for the reproduction of these spaces into consumer sites for tourists and the wealthy.

Why revitalize the historic center? The historic center of Mexico City houses the largest collection of colonial architecture not only in Mexico, but across Latin America, including the Metropolitan Cathedral, one of the largest cathedrals in Latin America and the world after the cathedral found in Sevilla, Spain. The streets, plazas, churches, and palaces are charged with national symbolism that represent the birth of the Mexican people. The symbolic and cultural value of the historic center is a major selling point for national and international tourism—with the goal of city officials and investors of turning this cultural value into tangible profits.

Mexico City is taking the cue from cities around the world to reinvent the historic center with the goal of catapulting the city into Global City status. The program intends to improve the city's position within the global network by creating infrastructure to encourage global investment in real estate, tourism, businesses and corporate headquarters, the professionalization of labor (i.e., the relocation of street vendors into permanent stalls), and high-end recreation and consumption. The program aims to regulate directly who has access to public urban space and housing and how urban space is produced and consumed. This plan is in direct conflict with street vendors, Santa Muerte shrine keepers, and worshipers' conceptualization of the production and utilization of urban space and their right to the city. The social geographies that the city bureaucrats and private investors seek to produce collide with those that are imagined and reproduced by the working-poor street vendors and worshipers.

In 2003, Andrés Manuel López Obrador (AMLO), then mayor of Mexico City who was elected President of Mexico in 2018, announced that the public–private "recovery" of the historic center would be accomplished in conjuncture with Grupo Carso, a transnational corporation run by Carlos Slim. The plan is still in progress.

Over the last 20 years, 4.5 billion pesos have been spent in rehabilitating 34 square blocks, including the Alameda Central, one of the few green areas in the historic center flanked by the Fine Arts Theatre and a museum housing a Diego Rivera mural (Infobae, 2021). The revitalization of these 34 blocks, or Perimeter A, consists of less than half of the historic center. The "rescued" section runs from Jesus Maria Street, two blocks east of the Zocalo behind the National Palace, to Balderas Street, one block west of the Almeda, which extends behind the Museum of the Diego Rivera Mural, *Dream of a Sunday Afternoon in Alameda Central.* And to Belisario Dominguez Street three block north of the Zocalo behind the National Cathedral and two blocks south of the Zocalo to Republic of Uruguay Street. Within Perimeter A, two of the main east–west arteries have been pedestrianized, Cinco de Mayo and Francisco Madero (see Figure 2.10). The 17th- and 18th-century palaces have been restored. And yes, the street vendors have been removed through punitive measures.

In addition to the *Plan de Rescate*, following a report released by the international security consultancy Giuliani Partners, city officials enacted the Civic Culture Law of the Federal District (*Ley de Cultura Cívica del Distrito Federal*, LCC) in 2004 to enhance the "quality of life" in the city. Even though the law was presented as a means for promoting "harmonious cohabitation" in the city, it specifies 43 administrative infractions—most related to informal activities and the improper use of urban spaces—that is, street vending—that are punishable by up to 36 hours of confinement and fines of between 1- and 30-days' earnings at minimum wage. The Civic Culture Law of the Federal District was implemented by authorities in an effort to remove informal street vendors from the historic center. In the month of October 2007 there were nearly 15,000 arrests.

The implementation of the Civic Culture Law of the Federal District for the daily reality of street vendors in Perimeter A involved harassment and arrests. From the 2006 to 2011 alone, the number of detainees who were brought before the civic judges (*juzgados civícos*) in charge of punishing Civic Culture Law-related administrative offenses nearly tripled from 49,205 to 134,731. About half of these cases were related to arrests in Cuauhtémoc, the borough that is home to the historic district. In that borough alone, civic judges heard 56,719 cases regarding Civic Culture Law-related infractions, mostly the "obstruction" of public spaces—again, involving street vending—from August 2010 to September 2011.

In conjunction with the punitive implementation of the Civic Culture Law, AMLO's administration used Carlos Slim's money for additional security investments. To further control the urban space, the administration installed CCTVs and created and trained a special historic center police force with a central command center that can respond immediately to any security problem or street vending. After much investment the plan has successfully converted Perimeter A into a site of high consumption with appeal to global consumers—be they Mexicans, French, or gringos, while simultaneously violating the civil rights of its citizens (Müller, 2021).

Perimeter B has not seen the same level of "recovery," but not for lack of investment. For example, the city rehabilitated and converted a 17th-century palace into the Museum of Light (Figure 2.11). This university-run museum dedicated to educating the public about the science of refracting light was opened in 2005. However, the city failed in removing the street venders from this area. The large crowds surrounding the street vendors discouraged tourists from visiting the museum. By 2012 the museum became inactive.

How have the street vendors fared in Tepito just north of the Zocalo? Tepito has always been a neighborhood of resistance (Ureste, 2012). Situated just eight blocks from the Zocalo, the epicenter of political and religious power, the neighborhood has been able to maintain its identity. Street vending, use of the folkloric *albur*, the rise of the syncretic form of Santa Muerte adoration, and vernacular food practices: These factors helped Tepito's residents mobilize their identities to facilitate the appropriation of urban space to negotiate economic shifts in the national and global economy through the CET. This strong sense of place—identifying and belonging to the historic center, to Tepito—combined

**FIGURE 2.11** Street vendors remain at site of the Museum of Light. Photo by David M. Walker.

with livelihood strategies always already occurring in the street—has also resulted in residents mobilizing to protest the further removal of street vendors (Milenio, 2020). Many incidents of blocking roads that clog the transportation arteries and prevent the mobilization of police, combined with violent outbursts of the protesters against authorities, have deterred city officials from successfully containing street vending and the informal activity in this space. Street vending and Santa Muerte worship have not been removed from Tepito. Attempts to "normalize" economic activities in the historic center by relocating street vendors into markets and warehouses in working-class neighborhoods, which would allow the authorities to charge for permits and sales tax and to regulate transactions have yet to occur. Through the informal economy combined with Santa Muerte worship in public altars and shrines found throughout the neighborhood—the coalescing factors that mobilize Tepito residents to appropriate urban space—they have reproduced the social urban fabric of this locale and have embedded themselves into these spaces. These strategies have thwarted the neoliberal "*rescate*" program for reconfiguring the social geographies of Tepito.

## The Appropriation or Rebranding of Tepito

Professional Tepito entrepreneurs, such as Alfonso Hernandez, who created the CET and *Arte Acá* organizations, are also learning neoliberal rebranding strategies and the importance of spectacle in reproducing urban space (Ibarra, 2023). Taking a cue from city officials who created the Spring Festivals in the historic center to promote the newly rehabilitated spaces and attract consumers, *Arte Acá*, a cultural group in Tepito dedicated to cultural activities, especially the creation of murals, organizes a neighborhood festival for Tepito. With sponsors such as the secretary of culture and the National Filmmakers Association, the festival's goals are to "take back (*retomar*) the camaraderie, collegiality, and hold tertulias, or social gatherings where informal discussions or symposiums occur. The main activities of *las tertulias*, which meet in the streets and housing complexes of the neighborhood, center on Tepito's importance in producing famous footballers, boxers, street vendors, *Alburistas* (people adept at speaking in *albures*), Santa Muerte shrines, and historical events, such as being the site for the last encounter between the

Aztec Kings Cuauhtémoc and Cuitláhuac. *Arte Acá* has recognized the importance of the quotidian social geographies embedded in the spaces of Tepito and has begun to offer tours of the neighborhood. Objectifying the social geographies of Tepito, *Arte Acá* calls these tours *Un "safari" por el barrio bravo de Tepito* (Ureste, 2012), to promote the perceived "exotic adventure" tourists will find on these safaris such as Santa Muerte shrines, street vendors, food and beverage stands, and colorful neighborhood characters.

The following section describes an event in Tepito called *Día de Los Muertos*. The description of this event provides an example of the production and appropriation of urban space in this neighborhood.

## The Appropriation of Urban Space through Street Vending, Worship, and Food: *Día de Los Muertos* in Tepito at Number 12 Alfarería Street

The main altar to La Santa Muerte is glowing with candles as people swarm to pay her tribute. The glass-incased shrine is filling up quickly with pesos and dollars, cigars, marijuana, crack cocaine, ornately printed prayers on laminated paper, and requests and messages scribbled on notepaper and napkins. The streets are teeming with revelers, many carrying small- to life-size statues of the deity, who are drinking, smoking, eating, setting off fireworks, and encouraging the musicians to play their favorite songs for *La Flaca* at 150 pesos each. The cacophony of sounds, the blend of smells from the gunpowder, food carts and stalls, beer, brandy and tequila, and marijuana and tobacco fill the air. There are *mariachis* playing horns and trumpets and plucking the *guitarrónes*, or large, fat guitars typical of the state of Jalisco. Gyrating bodies rapidly use mallets to strike *marimabas* that fill the night sky with songs from the Gulf Coast and the Isthmus of Tehantepec. Musicians playing large harps and *jaranas*, the small ukulele-style guitars from Veracruz, strum melodious songs in the *Jarocho* and *Huasteca* style. And the ever-present Norteño Trios play *narcocorridos*, the famous drug ballads that recount the stories of the daring bravado of young men and women avoiding the law and getting rich quick by "trampoline-ing" the product north for the consumption of the gringos, leaving abject poverty behind and giving back to the pueblo (community) in the style of Chapo Guzman.

Bustling crowds have come to pay honor to *La Flaca*. Vendors hawk candles, incense and sage brush for cleansing; others push carts laden with beer and *tapiche* (a fermented fruit beverage), and peddle small booklets containing prayers to La Santa Muerte. Worshipers come together around the food carts and stalls, making them serve as virtual moveable altars for her veneration. Beautiful green or red tamales, corn dough stuffed with pork in a green tomatillo or red chile de arbol sauce, wrapped and steamed in banana leaves, can be purchased for 10 pesos each. A plethora of tacos are available: *buches* (pork stomach), *lengua* (tongue), *tripa* (cow intestine), *suadero* (slow-stewed beef), *de canasta*, "basket" tacos in which the tacos are made ahead of time and then become soggy and, for the Mexican palate, more scrumptious, and *de cabeza* (delicious goat head tacos). Blue corn, which grows abundantly in the highland plateaus of Meso-America, produces airy, moist, and succulent tortillas that are filled with *quesillo*, a mozzarella-textured cow cheese, and *huitlacoche*, which literally translated from Nahuatl (the Aztec language still spoken by 2 million Mexicans) means the excrement of the gods. This delicious corn fungus is a delicacy in this part of Mexico (the blue corn quesadillas are my favorite).

The *Día de Los Muertos* celebration of La Santa Muerte functions as a triangulating device and as an identity-building mechanism among those who wish to protect Tepito for the worship of *La Flaca* (a nickname for La Santa Muerte) and preserve the informal livelihood strategies of the residents in the face of neoliberal policies. This festive occasion is a time of adoration, revelry, and a time and place for the exchange of goods and services, such as candles, prayer cards, and food and libations. Through the informal economies of worship, syncretized religion, and street vending, the people are producing and appropriating the urban spaces of Tepito. Here we witness the combination of alternative worshiping practices, the informal economy, and the survival tactics in contemporary neoliberal Mexico. This is syncretic México: *Mexicanos* who embrace *Mexicanidad* (Mexicanness) and worship their own vernacular deity.

## Neoliberal Success?

Mexico City receives many accolades for its culinary and artistic renaissance—the word of choice among travel and food writers—(Malkin,

2022) that has emerged in tandem with the revitalization of Perimeter A. In 2018, the World Design Organization named Mexico City its "World Design Capital." The Oscars success of Alfonso Cuarón's film *Roma*, named after the district in which it's set, has also added to the city's perceived cultural luster. The filming of the opening scene in the James Bond film *Specter*—set in the middle of a fictitious Day of the Dead parade in the historic center with a final helicopter-fight act in the Zocalo—motivated the country's tourism board to create an annual parade. "When this movie hit the big screen and was seen by millions and millions of people in 67 countries, that started to create expectations that we would have something," explained Lourdes Berho, CEO of the government's Mexico Tourism Board (Stevenson, 2019). Thus, the neoliberal revitalization and promotion of the historic center had gone global: It influenced art, and then art influenced the social geographies of the historic center. Part of Mexico City's cultural and tourism successes are linked to investments in the historic center.

The revitalization plan, the street vendors, and Santa Muerte worshipers have been successful at reproducing their distinct social geographies in these locales. Neither entity has completely achieved what they wanted or completely lost what they had. Street vendors have given up vending spaces in the historic center while maintaining Santa Muerte altars in Tepito. Government officials still struggle to contain the street vendors, and while the central government and Catholic Church discourage *Santa Muerte* worship, it has only increased in popularity.

Earlier in this chapter it was suggested that La Santa Muerte will eventually be beatified by the Catholic Church and be formally recognized, just as what occurred with the syncretic Virgen de Guadalupe. As described earlier in "The Appropriation or Rebranding of Tepito," *Arte Acá* organizes Safari Tours of Tepito, and the annual neighborhood festival effectively rebrands Tepito through the adaptation of neoliberal urban revitalization strategies. Do these activities signify that street vending, *albures*, street food, and Santa Muerte worship will also become accepted and a selling point of Tepito and eventually promoted and recognized by the state tourism board, effectively intertwining what is accepted and attractive and what is taboo (see Figure 2.12 for an example)? The Safari Tours and Fiesta del Barrio demonstrate the pervasiveness of neoliberal practices.

FIGURE 2.12 A shrine where La Santa Muerte is venerated with Jesus Christ. Photo by David M. Walker.

## Conclusion

There are multiple ways that cities are imagined and reproduced. Conflicts over how urban space should be utilized resonate from Tepito, the entire Historic District, and beyond. The contestation over the historic center provides an example of how neoliberalization impacts cities across the globe, and for the purpose of this chapter, we can learn how street vendors and Santa Muerte worshipers use urban space to maintain their way of life.

In this chapter I described how cross-scalar economic and cultural policies and practices are grounded in a local place. The chapter tells the story of the rise of neoliberalism, the growth of the informal economy and street vending, Santa Muerte worship in the historic center of Mexico City, and the successes of the city's revitalization program. As Doña Queta explained "We [Mexican People] are going through some very tough times, economically, morally, and physically; we need to grab hold a little of everything in order to make it, to stand up and keep living until death takes us . . . La Flaca is not evil; she is not good or bad. She only negotiates with people. The things that people seek her collaboration with may be interpreted as wrong or correct . . ." (D. Queta, personal communication, 2017). The government of Mexico City partially got what it wanted, so far: the "recovery" of nearly half the historic district. And people living in Tepito were able to partially keep, "to grab hold [of] a little of everything in order to make it" what they wanted: to maintain the informal economy and La Santa Muerte adoration in Tepito but not across Perimeter A of the historic center. Finally, in this chapter I tell one of the many stories of how a Mexican family is engaging with life in contemporary neoliberal Mexico.

## DISCUSSION QUESTIONS

1 Who or what is La Santa Muerte? Describe how she represents religious and cultural syncretism and explain why her popularity has increased in Mexico since the 1980s.

2 Reflecting on the description of neoliberalism and the neoliberalism of space discussed in the chapter, consider how neoliberalism in your community or country has transformed a place or space (i.e., private highways, sponsored sports stadiums, neglected and dilapidated bridges and roadways, privatized water systems, schools and for-profit colleges and universities?)?

3 Describe how the informal economy takes place in the public sphere on the streets and plazas of Tepito and in the historic center. Now, brainstorm three examples of the informal economy in your community or country (excluding illegal activities such as drug dealing or smuggling). What are they? Where do they take place? How do they compare with or differ from the strategies of street vendors in Mexico City? Finally explain why the federal, state, and municipal governments in your country of origin would want to prohibit informal economic activities and livelihood strategies and how those policies compare and contrast with the government of Mexico City?

4 Explain how and from where Alfonso Hernandez, who created the CET and *Arte Acá* organizations, learned neoliberal rebranding strategies and the importance of the spectacle in reproducing urban space to promote the newly rehabilitated spaces and attract consumers to Tepito. What type of events did these organizations create and how did they brand Tepito tours?

5 Explain how the emergence and eventual acceptance and embracing of the Virgin of Guadalupe by the Catholic Church is similar to and different from the rise in popularity of La Santa Muerte.

6 The informal economy and street vending are not new to Mexico City, as these phenomena can be traced to the pre-Colombian era through the colonial period. Explain how and why street vending increased beginning in the 1980s?

7 What are the reasons why we witnessed the decline of the middle class beginning in 1982 in Mexico City?

8 How are ISI and urbanization mutually constitutive processes? Why did Mexico City grow so rapidly from 1960 through 1985 during the implementation of ISI policies?

## SUGGESTED RESOURCES

To learn more about La Santa Muerte and the politics of gentrification in the historic center of Mexico City, see:

Walker, D. (2008). *Gentrification moves to the Global South: An analysis of the Programa De Rescate, a neoliberal urban policy in Mexico City's Centro Histórico* (Doctoral dissertation). Retrieved from *https://uknowledge.uky.edu/gradschool_diss/654.*

For an examination the character of Mexico City and how the Mexican national character shaped and was shaped by the capital city, see:

Johns, M. (1997). *The city of Mexico in the Age of Diaz*. Austin, TX: University of Texas Press.

For a 21st-century understanding of and insight into class dynamics and urban space in Mexico City, see:

*Amores Perros*. (2000). Directed by Alejandro González Iñárritu and written by Guillermo Arriaga.

For an accessible and comprehensive history of Mexico City, see:

Kandell, J. (1988). *La capital: The biography of Mexico City*. New York: Random House.

### REFERENCES

Adeath, C., and Kristensen, R. (2007). *La Muerte de tu lado*. Casa Vecina Mexico: Coleccción Libros de la Meseta.

Anaya, A. (2014). The albur and refran as tropes for identity construction in Mexico. *Bulletin of the Transylvania University of Brasov, 7*(56), 2.

Aridjis, H. (2017). *La Santa Muerte*. Barcelona, Spain: Debolsillo.

Díaz del Castillo, B., Smith, J. B., and Humphrey, T. (2012). *The true history of the conquest of New Spain*. Indianapolis: Hackett.

Ibarra, J. (2023, June 13). Durante Un Mes, Tepito Celebrará Su Festival de Barrio. *La Jornada*.

Infobae. (2021). El día que Carlos Slim decidió rescatar El Centro Histórico de la CDMX. Retrieved May 7, 2021, from *http://infobae.com/america/mexico/2021/05/07/el-dia-que-carlos-slim-decidio-rescatar-el-centro-historico-de-la-cdmx*.

Malkin, E. (2022, November 17). 36 hours in Mexico City. *New York Times*. Available at *www.nytimes.com/interactive/2022/11/17/travel/things-to-do-mexico-city.html?searchResultPosition=1*.

Matas Rosas, F. (2010). *Oceano*. Tepito, Mexico: EL Barrio Bravo.

Milenio, D. (2020). Bloquean Eje 1 Norte y Eje Central, Comerciantes de Tepito. Retrieved July 3, 2020, from *http://milenio.com/videos/politica/comunidad/comerciantes-tepito-bloquean-eje-1-norte-eje-central*.

Müller, M. M. (2021). *The punitive city privatized policing and protection in neoliberal Mexico*. London: Zed Books.

Paz, O. (1961). *The labyrinth of solitude: Life and thought in Mexico*. New York: Grove Press/Evergreen.

Peck, J., and Tickell, A. (2002). Neoliberalizing space. *Antipode, 34*(3), 380–403.

Smith, N. (1987). Gentrification and the rent gap. *Annals of the Association of American Geographers, 77*(3), 462–465.

Stevenson, M. (2019, November 5). Mexico City's first day of the dead parade inspired by James Bond film "Spectre." *The Hollywood Reporter*. Available at *http://hollywoodreporter.com/news/general-news/mexico-citys-first-day-dead-parade-inspired-by-james-bond-film-spectre-942883*.

Ureste, M. (2012). Animal politico. Retrieved June 18, 2024, from *http://animalpolitico.com/sociedad/un-safari-por-el-barrio-bravo-de-tepito*.

Walker, D. M. (2008). *Gentrification moves to the global south: An analysis of the Programa de Rescate, a neoliberal urban policy in Me?xico City's Centro Histórico*. Unpublished manuscript, University of Kentucky.

Wolf, E. (1958). The Virgin of Guadalupe: A Mexican national symbol. *Journal of American Folklore, 71*, 34–39.

# CHAPTER 3

# Identity and Place

## *The Social Geographies in the Costa Chica of Guerrero and Oaxaca*

### Singing the National Anthem in Mexico City

Many *Costeños* do not realize that they are anything other than Mexican until they leave the Costa Chica. *Costeños* share many anecdotes about traveling to Mexico City for work or to attend the university. In Mexico City they are confronted with being categorized as negro or *pardo* (a term used in Mexican Spanish to describe people of African descent).

*Costeños* describe how when riding the metro or buses or walking the streets in Mexico City they are frequently stopped by the police and questioned if they are Mexican. The authorities ask them for identification. Upon showing their national identity cards, student I.D. cards, or even passports, the police will continue to question them because they do not believe that they are *puros Mexicanos*: believing they must come from Cuba and be undocumented in Mexico, searching for work or earning a living as smugglers or prostitutes. This humiliating treatment of the *Costeños* by the authorities often accelerates, with the police demanding that the *Costeños*–Afromexicanos sing the national anthem to prove they are truly citizens of the Mexican state—can you imagine U.S. residents having to prove their citizenship by singing the national anthem (yes, Canadians all know and can sing *O Canada!*, but fewer

English people know all the lyrics to *God Save the Queen*, but can hum the tune)! This abuse is an affront to the *Costeños* living, working, and studying in Mexico City. The demand by the authorities, who represent the state hegemonic power structure, and therefore the right to legitimate violence, that *Costeños* sing the national anthem demonstrates how the Black populations across México do not exist in the eyes of the state or in the national imagination of the *Pueblo Mexicano*, or Mexican people. Most Mexicans simply don't know that Black Mexicans exist: Afromexicanos are born and raised in Mexico, are Mexican citizens, and have been in México since the time of *el disencuentro (or colonization). Most Mexicans don't know that the important historical* actors Vincente Guerrero, Miguel Hidalgo, José Morelos, and others who helped shatter the yoke of Spanish colonialism in order to forge the modern Mexican state were Afro-descendants. Why? Because Black history has been erased from the national memory—it is not in Mexican textbooks or reflected in the Mexican national anthem (Mansferrer Leon, 2018).

To study the African diaspora is to study geography. Black feminist geographer Katherine McKittrick reinforces this connection through her affirmation that "Black matters are spatial matters," and that Black communities across the diaspora "struggle with discourses that erase and despatialize their sense of place" (2006, p. xiii). Such has been the case for *Costeños* who travel across the Mexican republic, as indicated in the previous anecdote. This is also the case for the vernacular region of Costa Chica: It has been despatialized through the erasure by the federal government that Black African spaces exist in Mexico.

Afromexicanos have resided in the villages and towns of the Costa Chica for the last 400 years. Within the last 20 years residents have participated in a movement occurring across Latin America that consists of a rise in Black awareness, Black pride, and Black power to be recognized by the state and to be counted on the official census in order to be counted as full citizens. This Black social movement is closely linked to and even modeled on autonomous Indigenous movements that seek to be recognized by the federal government and included as a racial category on the census. Additionally, Costa Chica residents (*Costeños*) are mobilizing their Black political identity and consciousness as it relates to the construction of region and place to make formal demands for an official regional recognition of the Costa Chica by the national and state Mexican governments with the goal of receiving federal and state development aid and control of natural resources.

## Where Is the Costa Chica Today?

The Costa Chica is located along the foothills and Pacific Plains that extend through the wetlands, bays, lakes, and lagoons between the Sierra Madre del Sur Mountain ranges and the Pacific Ocean, and from Acapulco, Guerrero, to Puerto Nuevo in the Pochutla district of Oaxaca (Figure 3.1) Throughout the foothills and low-lying tropical zone, residents grow coffee, mango, papaya, peanuts, marijuana, and tobacco and subsistence crops such as corn, beans, squash, and chiles—the classic *milpa* production of peasants across México. They also fish (Figures 3.2 and 3.3), raise chickens and pigs, and have a small but culturally important cattle industry. Dozens of foothill and coastal villages house many Afromexicano Costa Chica residents, or *Costeños*. Indigenous Mixtecas reside in the villages of the Sierra Madre del Sur.

The towns of Cuajinicuilapa, Guerrero, and Pinotepa Nacional, Oaxaca, each roughly 30 km from the coast, represent the major trading and administrative centers in the Costa Chica. The markets in these towns teem with Afromexicanos selling and buying goods cheek by jowl with Indigenous Mixtecas. Indeed, market experiences and personal and economic interactions function as a signifier of the ethnic and racial differentiation that fuels identity construction for both groups (S. Hall, personal communication, 1996). Although the linguistic, ethnic, and racial differences distinguish the Afromexicano and Indigenous Mixtecas, they are unified in their shared marginalized positionality and mutual plight to gain control of their resources from the primarily mestizo and European-white elite power structure at the local, state, and national level.

The Costa Chica is a vernacular region, in that it is known colloquially or locally by this name, and is not on any official tourism or development maps. The region includes not only the coastal villages, but also the two main cities and additional villages found throughout the plains and foothills just described.

## Regions in Oaxaca and Guerrero

At present, the federal government of México City and the state government of Oaxaca officially recognize seven distinctive regions in the state of Oaxaca: Istmo, Valles Centrales, la Cañada, Papaloapan, Sierra

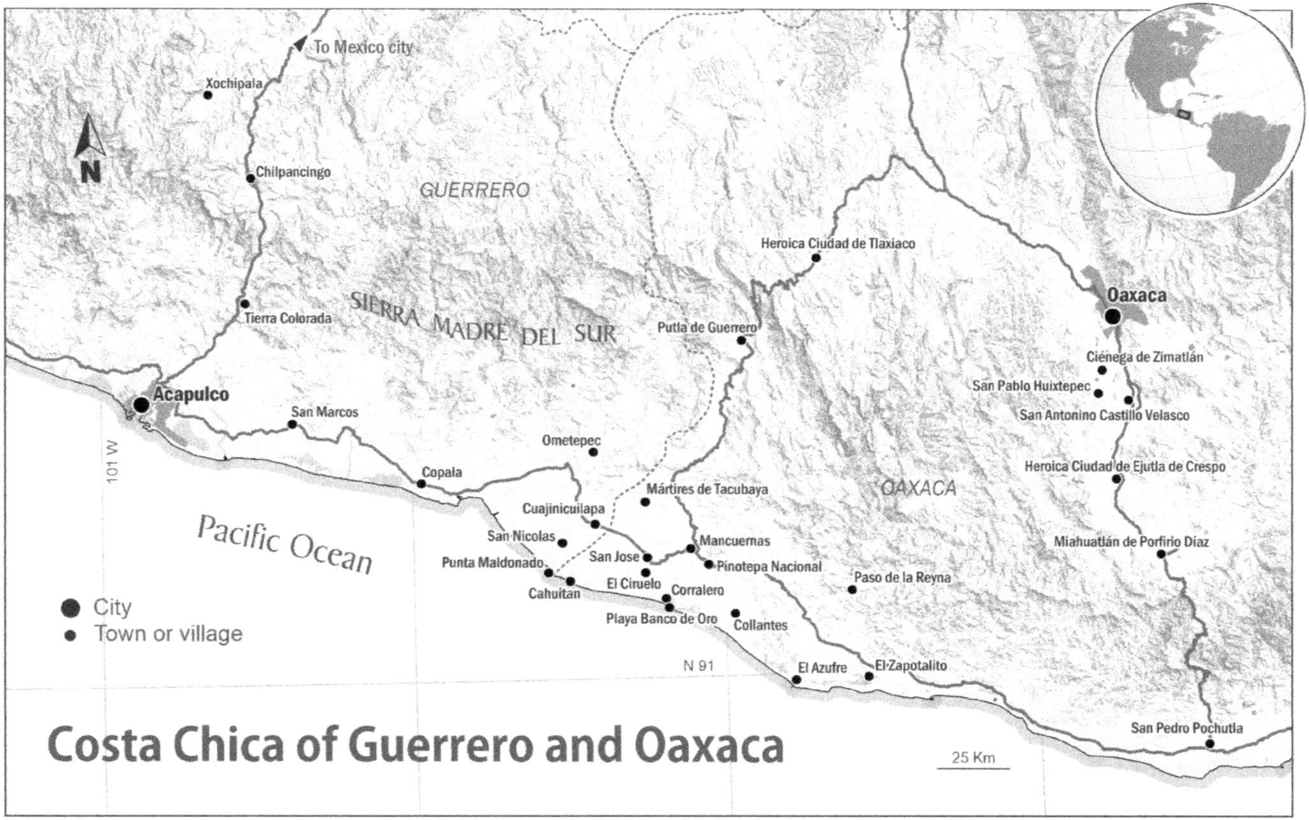


**FIGURE 3.1** A map showing the vernacular region of the Costa Chica located between the Sierra Madre del Sur Mountain ranges and the Pacific Ocean, and from Acapulco, Guerrero, to the Pochutla district of Oaxaca.

FIGURE 3.2 Casting a net for shrimp in Corralero Lagoon, in Corralero, an Afro-Mexican town on the Pacific Coast of Oaxaca, Costa Chica. Photo by David M. Walker.

FIGURE 3.3 Sun-drying fish in Corralero to be sold at the market in Pinotepa Nacional. Photo by David M. Walker.

Norte, la Mixteca Alta, and la Mixteca Baja. These seven regions are distinguished by the representation of their Indigenous idiosyncrasies. The state of Guerrero consists of the La Costa and La Sierra regions. Currently, the vernacular region of the Costa Chica spans the Mixteca Baja region of Oaxaca and the La Costa region of Guerrero.

Official regional recognition facilitates tourism promotion and aids local officials in their request for development aid for the respective region. This is especially true in Oaxaca, where the district cultures of Indigenous groups have piqued national and international interest, leading to an increase in tourism. Without official regional recognition, it is impossible to exercise autonomy over the natural resources found in the region, primarily in agriculture, aquaculture, and fishing, and to gain access to development funds to build roads, schools, hospitals, and a tourism infrastructure.

The following section describes the important contributions that Africans, a large part of the population, have made to the formation of Mexico. Africans have always been part of Mexico since its conception. Nevertheless, in Mexico and abroad, few people are aware of this Black history.

## Africans in Mexico

During the colonial period, over 300,000 persons of African descent entered New Spain (Aguirre Beltrán, 1972). Black conquistadors rode into Tenochtitlan with Cortez, and Estevanico, a Black explorer, made the land-route journey from Florida to Mexico City from the southeastern and southwestern part of the United States with the explorer and conquistador Panfilo de Narváez (Vinson and Vaughn, 2004). Afromexicans participated in nearly every historical turn in México's trajectory as a country—from Afromexican Vicente Ramón Guerrero, one of the leading revolutionary generals of the Mexican War of Independence who became Mexico's second president (yes, Mexico's second president, whose father was Afromexicano and mother Mestiza), to Miguel Hidalgo, whose battle cry "Grito de Dolores" called the populace to arms to end the hegemony of Spanish colonialism. Indeed, Blacks accounted for the majority of the troops in Mexico's 1810–1821 War of Independence (Ramsay, 2004). Guerrero and Hidalgo are two historical figures who are included in textbooks and whose stories are taught to

Mexican students across the republic. However, their identities as leaders of African descent are erased from the official accounts. And what of the women of African descent who surely participated alongside men in the creation of the Mexican state? Their identities are not just erased, their intersectional voices are not even heard in the official state-sanctioned stories.

According to the 1793 census, 270 years after Cortez marched into Tenochtitlan to lay claim to the Aztec Empire in the name of Spain and Catholicism and to get rich, Europeans, or *peninsulares* as they were called, numbered 7,900, or just 0.2 percent of the population, and Africanos were 0.1 percent with 6,100 people. Incredibly, Afromestizos (Black with Indigenous blood) made up 10 percent of the population, or 370,000 people; Indomestizos made up 11.2 percent, or 420,000 people; and Euromestizos made up 1 percent, or 680,00 people. In 1793 Indigenous people held a clear majority with 61 percent of the population numbered at 2.3 million people in New Spain, exemplifying the demographic rebound that they experienced in Mexico after the demographic collapse that occurred with the arrival of Europeans 270 years earlier (Mann, 2005).

On the eve of Mexican independence in 1810, the Afromestizo population would nearly double to 630,000 residents, whereas the Africanos population increased by just 4,000 during the same time frame (see Table 3.1). Unfortunately, there are no National Institute of Statistics and Geography data about gender distribution for these dates.

According to the 2015 midnational census, nearly 1.4 million people self-identified as being of African descent (INEGI, 2019). As a percentage of the total population across Mexico, Guerrero (6.5 percent), Oaxaca (4.9 percent), and Veracruz (3.3 percent) represented the states with the highest proportion of people who identify as Afromexicano. In 2020 slightly over 50 perçent of the 2.6 million people who identified as Afromexicano were concentrated in six states, including Guerrero (300,000) and Oaxaca (200,000).

These historical and contemporary statistics reveal the rapid early growth of African populations in Mexico (1600–1780), with an increase in the number of *mestizaje* resulting in a quicker growth of Afromestizos compared to Africans from 1789–1810.

Through human–environment interaction, Africans would form the population found across the sugar plantations of the Gulf Coast and the mountains of Veracruz, the mines of Taxco and San Luis Potosi, the coffee plantations in the Putla region, and, important for this study, the

**TABLE 3.1. Population Changes in the Viceroyalty of New Spain, 1792–1810**

| Casta | Inhabitants 1793 | Percent 1793 | Inhabitants 1810 | Percent 1810 |
|---|---|---|---|---|
| Europeos | 7,904 | 0.2 | 18,000 | 0.2 |
| Africanos | 6,100 | 0.1 | 10,000 | 0.1 |
| Indígenas | 2,319,741 | 61.0 | 3,676,281 | 60.0 |
| Euromestizos | 677,458 | 17.8 | 1,092,367 | 17.9 |
| Afromestizos | 369,790 | 9.6 | 624,461 | 10.1 |
| Indomestizos | 418,568 | 11.2 | 704,245 | 11.5 |
| Total | 3,799,561 | 100.0 | 6,122,354 | 100.0 |

*Note.* Table constructed by David M. Walker.

Costa Chica. Guerrero self-identifies as 8.6 percent Afromexicano and Oaxaca as 4.7 percent Afromexicano (INEGI, 2019), with the largest Afromexicano populations living in municipalities found in the Costa Chica. This next section delineates the different ways that the African descendants arrived in the Costa Chica and how their arrival has served to discursively embed the *Costeño*–Afromexicano identity into this place.

**VIGNETTE: *Geography Matters: Human–Environmental–Diasporic Interactions and Environmental Determinism***

The correlation between the spread of African populations across the western hemisphere and human–environmental interaction demonstrates how societies are dependent upon nature to make a living and are also extremely capable of modifying the natural environment: Geography matters. This is not the same as environmental determinism. We can appreciate this spatially by considering the map in Figure 3.4. The lands dedicated to plantation economies across the volcanic soil islands of the Caribbean; in the southeast of the United States; and in northeastern and southeastern Brazil are environments where sugar, coffee, and rice and the important textile crop cotton are well adapted to cultivation.

The Caribbean tropical zone is the first area settled by Spanish explorers, missionaries, and conquistadors upon arriving in the Americas. Their arrival did not bode well for the Indigenous communities

*(continued)*

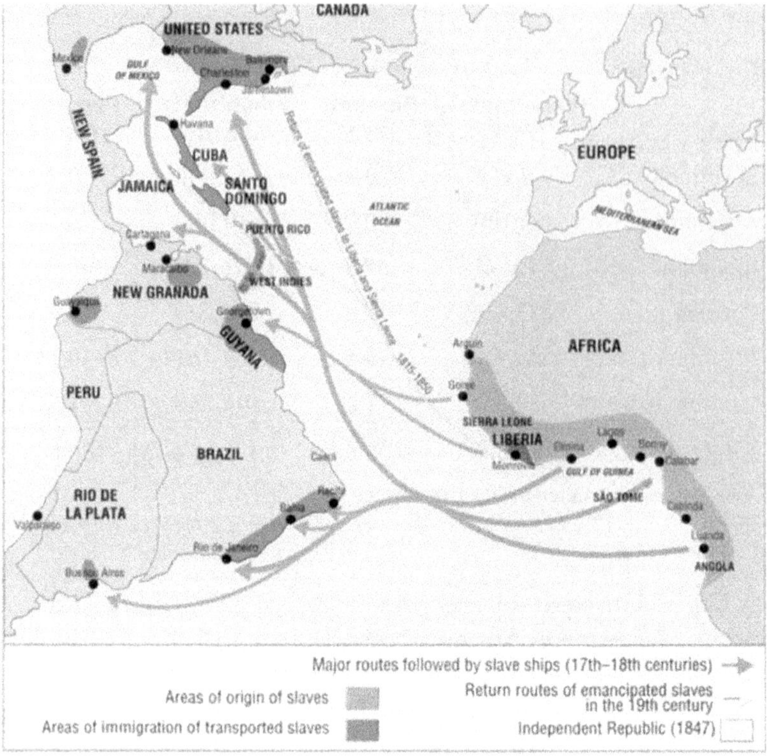


**FIGURE 3.4** Map showing African Diaspora across the Americas. From Marston et al. (2017). Copyright © 2017 Pearson. Reprinted by permission.

found across the region. The humid tropics are an insalubrious climate in which germs and infection spread quickly. This zone witnessed the beginnings of the New World Indigenous genocide and the European transatlantic slave trade, where on the island of Hispaniola (modern-day Haiti and the Dominican Republic) the first Africans came as slaves to the Americas in 1502 at the hands of Europeans.

From the 16th through the 19th century, the plantation economies increased production to satisfy the appetites of Europe and help fund the Industrial Revolution through profits made in the European Trans-Atlantic Slave Trade and in sugar, coffee, rice, and cotton production (Inikori, 2020). More than 12.5 million Africans were enslaved and brought to labor across the plantations and elsewhere, leaving an indelible mark on the economies and cultures of the Americas. Although there are now African-descendent populations living outside of these plantation zones, these regions still house the largest Black communities in the Americas, thus demonstrating how geography matters in relation to human–environmental interactions.

According to the theory of environmental determinism, which peaked in the late 19th and early 20th centuries and came to be generally accepted, the physical environment—and most specifically climate—dictated the sociocultural and economic aspects of society. Furthermore, proponents of the theory argued that the environment determined which cultures around the globe would be become socioeconomically and militarily dominant. The divergence of cultures that we experience around the globe today is not determined solely by a society's physical geographical surroundings. Levels of technology, systems of social organization, and ideas about what is true and right have no obvious relationship to environmental circumstances, because the cultural definition of resources, in principle, is that people, not the resources, determine what in the environment or nature is valued or not.

For example, "a rich deposit of uranium ore is of no value for a culture that does not use uranium. People may live in a place for generations, growing food, building houses; raising families . . . the tabulation of important landscape features in that place will not include uranium mines. Until there is a demand for it, an (uranium) ore deposit is not a resource" (Gersmehl, 2014, p. 114).

The environment does place certain limitations on the use of territory. Such limitations are not absolute, enduring restrictions, but are related to technologies, cost considerations, societal aspirations, and cultural norms. For example, it would be very costly to grow coffee in Minnesota, but not impossible with today's technology. The tropical and subtropical regions are prime areas for sugar, chocolate, coffee, and cotton production, but these zones are not limited to just these specific crops.

Europeans placed enormous value on exotic crops such as sugar, coffee, and cotton that could not be easily grown in Europe or accessed via the spice roads throughout the Middle East ruled by the Ottoman Empire. They made deliberate decisions to place a value on these natural resources; transport the crops to the Americas (an example of cultural and economic diffusion); cultivate the crops in large monocropping plantations, thereby forever transforming the landscapes of the Americas; capture, transport, and enslave skilled labor from Africa that had experience with and knowledge of farming in the tropics to till the fields and harvest the goods; and market these natural resources in Europe. Nothing about the environment dictated that Europeans were required

*(continued)*

to follow this model: that cotton could not have been grown in the southeastern United States by yeomen farmers in lieu of slave labor or that a global trade in cassava produced by European immigrants could have emerged from European's happenstance discovery of the Caribbean. Europeans in the Caribbean basin, the subtropical zones in the Americas, and elsewhere intentionally established systems of colonialism, slavery, monocropping, mercantilism, and imperialism because they had very specific goals and an agenda to enrich themselves with products not found in Europe and to use slave labor and eventually indentured labor to do so.

Interestingly, West Africans produced rice, sugar, cotton, and myriad other crops along the fertile river valleys found in West Africa for centuries. Indeed, at the time of European exploration in the late 15th and early 16th centuries, Africa and for that matter the Americas, too, were much healthier places to live than Europe (see Mann, 2005). The higher levels of well-being and the large populations found in Africa and the Americas at this time stemmed from healthy diets and hygiene, which prevented the spread in Africa of decimating pestilences such as the bubonic plague and others (we know that in the Americas the common cold did not exist, let alone measles and plague, but this has to do with the passage of Amerindians across the Bering Strait 35,000 years ago prior to the growth of large populations across Europe, as well as diet and hygiene—see Curry, 2019). The technological successes of West Africans in farming the tropics and of Amerindians in cultivating *milpas* and *chinampas*, for example (see Chapter 1), in similar climates meant that they did not need to travel to Europe to look for resources. In the case of Africa, the western African empires traded in salt, which they deemed an important resource (indeed, we can survive without coffee or sugar, but require salt in our diets).

These examples show us how different cultural societies (West Africans, Amerindians, and Europeans) value and exploit natural resources distinctively in similar environmental conditions, thus demonstrating that the theory of environmental determinism is one that does not hold true.

## Afromexicanos of La Costa Chica: A Legendary History

The director of the Afromexican Cultural Center in Pinotepa Nacional, Oaxaca, and the docents at the Afro-Mestizo Cultural Museum (Figure 3.5), in Cuajinicuilapa, Guerrero, promote an origin narrative of cimarron communities emerging in the Costa Chica during the early colonial period (personal communication, 2017). Cimarron communities are a spatial phenomenon that occurred in tandem with the formation of plantation economies in the Caribbean, Latin America, and the United States. *Cimarrones*, or *quilombos* in Brazil, *palenques* in Colombia, and maroon societies in the British Caribbean and the United States are townships formed by enslaved Africans fleeing enslavement on the plantations run by Europeans. Often Africans would align themselves with Indigenous communities as was the case in the Great Dismal Swamp in the United States and in the northeast of Brazil where the community of *palmares* grew to more than 30,000 people and lasted for 25 years.

Legend has it that Africans escaped from slave ships on route to the plantations along the Pacific coasts of Peru, Ecuador, and Colombia. The ships were caught in a storm that carried the vessels north and were shipwrecked off the Costa Chica where the Africans swam to freedom. This is a compelling story, as there does exist an important cultural connection between the west coast of South America and the Costa Chica

FIGURE 3.5 The Afro-Mestizo Cultural Museum in Cuajinicuilapa, Guerrero. Photo by David M. Walker.

that has played a major role in the formation of the region's social geographies. An example of the connection can be found in a musical style called the *chilena* (Stewart, 2013).

However, this connection took hold during the Gold Rush in California—nearly 250 years after the arrival of Africans to the Costa Chica—when ships would stop to barter and purchase water and rations and smuggle goods while traveling up and down the coast from San Francisco, California, to Chile and points in between. The syncretic musical genre of African origins called the *cueca* in Chile and Peru was introduced by sailors to the Costa Chica to the ports of Corralero, Collantes, Punto Maldonado, and surely Acapulco. Despite the *chilena's* transnational origins, over time it has come to be considered as a "constituent part of the large constellation of Mexican genres known as *son mexicano* . . . [*Chilena* music has] contributed to coalescing [Costa Chica] regional identity and became the emblematic performance genre of the Costa Chica" (Stewart, 2013, p. 72).

The cimarron origin narrative has been told and retold for centuries by the *Costeños* (Lewis, 2016). The *Costeños* proudly state that when the shipwrecked Africans came to shore, they were accepted by the Black communities that had already established communities who "owned this place . . . [as] here there were pure blacks, and they were the bosses . . . they accepted [the shipwrecked Africans] as their race [raza]" (Lewis, 2016, p. 8). The ubiquity of the shipwreck narratives provides strong evidence that cimarron communities emerged in the 17th century in the Costa Chica (Machado et al., 2019). In addition to the shipwrecked cimarrons, there is documentation that Africans arrived from a different route.

It appears that as early as the late 16th century "abandoned slaves" began to flee the silver and gold mines of Taxco, Guerrero, and begin to establish free communities in the Costa Chica (i.e., Negritud in Guerrero, Enciclopedia). This migration occurred when the viceroy of the time wanted to take control of the silver and gold miners' property. The owners chose to encourage the slaves to escape before that appropriation happened. The *cimarrones* from the mines joined with the slaves who fled the harsh conditions of the sugarcane fields in Morelos to establish well-documented communities in the Costa Chica, as witnessed and confirmed by a letter from Viceroy Luis de Velasco in 1606. "On the coast of the South Sea, near the port of Acapulco, there are three other ranches . . . of Blacks who say they will be more than 300 from where they do robberies (on the roads) of trains that carry and come to the discharge of Philippine ships." (General Archive of the Indies [Mexico], 2022, p. 27)

Thus, Blacks forged free communities in the Costa Chica prior to the famous uprising and formation of the "first free Black village" of the Americas called Yanga. Gaspar Yanga, a member of the royal family of Gabon, who was sold into slavery in Mexico, led a maroon colony of Africans in the highlands of Veracruz. He negotiated an agreement with the colonial government for self-rule, and the community of San Lorenzo de los Negros, or simply Yanga, was born.

Nevertheless, the shipwreck-origin narrative found in popular cultural representations and anecdotes in the Costa Chica should not be dismissed. The discursive power of this narrative unites the Black population of the Costa Chica.

The Centro Cultural Cimmaron in Ciruelo, Oaxaca, showcases murals drawn by a local artist, Aydée Rodriguez-Lopez. Central to the themes of the murals is the representation of a sinking slave ship that reputedly ran aground near the coastal village Punto Maldonaldo near Cuajinicuilapa, Guerrero (Weltman-Cisneros, 2013). One mural, Naufragio, depicts the now-freed Africans swimming to shore. In the foreground of the mural an African and a Spaniard emerge, both brandishing swords (see reference to access image, Rodríguez López, n.d.). The two images are not in conflict. The image represents "a narrative of dual self-empowerment, in which both figures contribute to the foundations of these Mexican communities in the Costa Chica . . . Both the European and the African are struggling for place in this scene." (Weltman-Cisneros, 2013, p. 132). This aspect of the mural "makes visible an alternative memory [the shipwreck origin of *Costeños*–Afromexicans] and an alternative archive of the presence and history of Afromexicans in the Costa Chica" (Weltman-Cisneros, 2013, p. 132).

Both actual and imagined communities in the Costa Chica were established by Afromexicanos from multiple sites, and as Rodriguez-Lopez's mural suggests coupled with the physical and facial features of Costa Chica inhabitants, the residents of these communities are a blend of African, European, and Indigenous peoples. Although historians, Spanish chronicles, or accounts at the time, or Mexican government archives do not document the shipwreck event recounted in anecdotes and depicted in local cultural practices, this does not mean that the narrative does not have the material effect of unifying Afromexicanos living in this place and contributing to the distinctive social geographies found in the Costa Chica.

The colonial demographic data (see Table 3.1) shows the large presence of Africans and Afromestizos in New Spain. What were the ramifications for the identities of people of African descent during the federalist period, (1810–1910) in the 20th century after the revolution, and how have these demographic trajectories impacted Afromexicanos today? It is important to learn about the path of Mexican racial categorization to understand why people of African descent in Mexico are fighting for representation today. The next section explains the origins of *la raza cósmica* in Mexico and the erasure of Blackness.

## *Mestizaje–La Raza Cósmica*

The Spanish colonial administration developed the *casta* or caste system to hierarchically categorize the Indigenous, African, and European mixed-race peoples found across the Americas, but especially in the Caribbean and New Spain. Related to Spanish ideas about purity of blood (with historical antecedents in the reconquest of Spain from the Moors), the *casta* system dictated more than socioracial classification. It influenced every aspect of life, including what profession you could have, where you could live (in towns and cities), and where you worshipped.

At the beginning of the Mexican wars for independence in 1810, Blacks made up 10 percent of the population, with free Blacks outnumbering the enslaved (see Table 3.1). The independence movement in México, when analyzed through the lens of the caste system, can be understood not only as a struggle for national independence, but as a social revolution (Vaughn, 2013). Indeed, Black communities in the coastal areas played important roles in battles against the Spanish. Their participation can be understood as a racial project, as defined by Omi and Winant (1994), in that " . . . the sociohistorical process by which racial categories are created, inhabited, transformed, and destroyed," are accomplished through "historically situated projects in which human bodies and social structures are represented and organized" (Omni and Winant, 1994, p. 69). The racial project aspect of this social transformation is further shown by the actions of Hidalgo who, in the first months of the conflict, declares the abolition of slavery and the elimination of the *casta* system. After Hidalgo's death, Morelos takes up the cause, banning slavery and *castas* and granting Indigenous rights. The wars represented more than achieving independence from Spain. For "darker" people, the independence wars represented a step toward equal opportunity and perhaps social integration. This social integration would be thwarted until the 21st century.

To end Spanish colonial hegemony and promote a distinct nationalism based on an alternative configuration of identity than the *castas*, Mexican political and intellectual elites mobilized the discourse of *mestizaje* as a means of reimagining Mexican citizenry. Independent Mexico passed laws prohibiting public officials from speaking disparagingly of any citizen's ethnic background and from using race in any government document and in any church records such as those for baptisms, marriages, and deaths. Under the guise of *mestizaje*, all residents would become Mexicans on the basis of a romanticized revision of indigeneity combined with Europeans and the erasure of Africans and their contributions.

After the Mexican Revolution (1910–1924), the discourse of *mestizaje* would be further mobilized and framed as a "*raza cósmica*," or a cosmic mixing of races that produced a "new race fashioned out of the treasures of all previous ones" (Vasconcelos, 1997, p. 40). A Mexican philosopher, pedagogue, and politician, José Vasconcelos manipulated the discourses of *mestizaje* and *la raza cósmica* as a tool to refashion Mexico as a unique cradle of the future of humanity. *La raza cósmica* would be an instrument of salvation similar to the trope of modernity to propel México into the future and eliminate any threat of material or discursive colonial designs and distance Mexico from its darker colonial heritage. The mobilization of this concept would impel Mexico to realize a final ethnic undertaking that integrated Europeans, Indigenous people, and Africans—*las tres raizes* (the three roots)—into an improved interracial *mestizaje*, where connection to the three founding roots would be eliminated and indistinguishable. Equally important for Vasconcelos and for Mexico, the formation of *la raza cósmica* categorized all Mexicans as one homogenized mosaic (Weltman-Cisneros, 2013), and differentiated the country from the Jim Crow-sanctioned institutional segregation and racism practiced in the United States.

Mexico's adherence to the cosmic race strategy continued until the last decade of the 20th century. This racially homogenizing concept had material effects on the lives of African descendants. (Indigenous communities have also been negatively impacted by this process; for a concise history of Mexico for those who would appreciate more historical context, see "Suggested Resources" in Chapter 1).

Only 65.5 percent of Afromexicanos have access to tap water, whereas 74 percent of all other Mexicans enjoy this service. In addition, 4.7 percent of Afromexicanos have no schooling, compared to 2.9 percent of other Mexicans, and 9.2 percent have incomplete elementary education, compared to 6.7 percent of other Mexicans. Afromexicanos also suffer from higher levels of illiteracy than the national average (Aguilar Rangel, 2022).

During colonialism, the Spanish recognized Blacks as slaves and lower-caste subjects. After independence, Mexico ended the practice of slavery and the *castas*, but influenced by the concept of *mestizaje* in which the *casta* system was changed to one based on class, Blacks were not counted and did not count. The postrevolutionary mobilization by Mexican elites of the *la raza cósmica* concept only made the plight of Afromexicanos more invisible, despite their contributions to the economy of México.

The next section describes theories about regions and identities and how they may overlap. This discussion is important not only in relation to how *Costeños* intentions are to be officially recognized as a marker for mobilizing their Black identity, but also in terms of how it spatially distinguishes their communities and potentially provides them access to power.

## Identity and Regions

Geographers believe that regions function as a way of seeing that which exists around us and as a means for organizing our thoughts about the world. Regions are based on socially constructed generalizations about the world, and although regional delimitation and representation are artifactual, they are by no means purely fictitious. It is important to recognize that regions are not static and that, much like place, space, culture, and identity, they are always in the process of becoming. The organization of space into regions functions as a medium for and an outcome of social practices and relations of power that are operative at multiple spatial and temporal scales, given that this study of the construction of the Costa Chica as an officially recognized region is contingent upon theorizations of place at the local, state, and national scales.

Theorists consider the production of identity in relation to psychic and social processes to be inextricably linked to the politics of representation. Identity is deployed not only as a critical tool useful in the analysis of cultural forms, but can also be utilized as a mechanism for confronting conflicts and mobilizing resistance by subaltern subjects who are subordinated by hegemonic processes related to gender, class, and race to gain access to and produce spaces and/or regions of representation.

The politics of identity are open to whoever wants to participate. Indeed, Afromexicanos in the Costa Chica receive support from outside the region, as witnessed by feminist Senator Susana Harp's political and financial support of the Afromexicanos struggle in the Costa Chica (Diariomo, 2018). The participation of Afromexicano NGOs, Indigenous groups, and women's and student groups combined with the broad acceptance of who can and does identify as Afromexicano (and when and where they choose to do so) also testify to the inclusivity that reinforces the politics of identity in relation to Afromexicano identity formation and official regional recognition of this locale. Whereas identity politics tends to be tied to a given identity (Mignolo, 2009), which prohibits the pluriversal construction of what and who you are, and when and where you show who you are, the concept surrounding the politics of identity is welcoming and inclusive.

The politics of identity function to mobilize new individual and collective possibilities. Conscious assertions of identity can be mobilized as a strategy of resistance by those dispossessed by the legacy of colonialism, patriarchy, slavery, and empire—as is the case for the Afromexicanos in the Costa Chica.

In this sense, the mobilization of the politics of identity plays a key role in the politics of location and the construction of place and region. The agency to form their own region in tandem with the assemblage of politics of identity represents to Costa Chica residents, as Stuart Hall puts it, not just a question of "who we are," or "where we have come from" so much as "what we might become, how we have been represented and how that bears on how we might represent ourselves" (Hall and du Gay, 1996, p. 4).

The next section describes the shift away from purely constructing a Mexican national identity in relation to *mestizaje* and *la raza cósmica* to a pluriethnic understanding of indigeneity. It is important to understand these changes in indigeneity as they relate to the creation of Indigenous autonomous spaces because these processes played an important role in educating Afromexicanos about how to mobilize civil-society organizations to work in tandem with state, national, and international groups.

## Indigeneity and the Emergent Afromexicano Civil Society

Under the far-reaching devolutionary reforms instituted during the neoliberal *sexenio*, or term of office, of Carlos Salinas de Gortari (1988–1994), the discussion of the significance of 1992 as marking 500 years of discovery and conquest, or 500 years of genocide, led to a constitutional recognition of Indigenous peoples and their rights. For example, in Oaxaca, where more than half of the vernacular Costa Chica region is located, these shifts led to changes to the electoral code, which in 1995 recognized traditional communal electoral practices, called *usos y costumbres* (Walker et al., 2007). *Usos y costumbres* are laws that govern everything in the municipality from agrarian reform to water rights and legally provide political autonomy to Indigenous communities. This recognition of traditional decision-making procedures outside of the political party system was overlaid by sweeping constitutional changes in favor of Indigenous communities and peoples (Oaxaca State Legislature, 1998a, 1998b; Rodriguez-Pose and Gill, 2004). As a result, roughly four-fifths of Oaxaca's 570 municipalities select local authorities based on local customs legally sanctioned under the *usos y costumbres* system and are considered to be autonomous Indigenous municipalities.

The strengthening of autonomy via the recognition of *usos y costumbres* occurred during a time when political claims on the basis of indigeneity had been powerfully articulated in México and elsewhere (Walker et al., 2007), whether in terms of social or economic justice or human rights (Oaxaca State Government, 2020). Nowhere was this push for autonomy made more apparent than in the southern state of Chiapas that borders Oaxaca, when, in January 1994, just as NAFTA went into effect, the Zapatista Army of National Liberation or EZLN seized control of parts of the state. The Zapatista rebellion remains the most visible, widespread, and long-standing postrevolution struggle by Indigenous groups in Mexico. The Zapatista uprising had the effect of catalyzing "politics of recognition" at the federal and state levels in México (Muñoz, 2004; Esteva, 2005). The politics of recognition sharpened in Mexico, eventually leading to political reforms that formally recognize cultural diversity and Indigenous peoples' rights. The recognition of cultural diversity at the federal level in 1994 produced unambiguously positive repercussions for Indigenous peoples, but none for other marginalized ethnic groups such as the Afromexicanos of the Costa Chica or the sizable population of Mexicans of Japanese and Chinese descent who live in Sonora and Baja California (Chao Romero, 2010; Hooker, 2005).

The close proximity of the Afromexicano and Indigenous Mixtecas allowed for professionalized *Costeños*–Afromexicanos to observe how Indigenous communities benefitted from their newly won autonomy. The shifts in Indigenous legal spaces, the racial awakening of *Costeño* migrants returning from the United States, where they experienced intersectional racism by being Black and Latino, combined with Black consciousness rising across Latin America, and the formation of Black NGOs working to demand Black rights from across the Caribbean to Columbia and Brazil, correlated with the mobilization of Black politics of identity in the Costa Chica.

The next section begins with a description of how Afromexicanos first came to work with Indigenous communities in 1992 and their work together was followed 5 years later by the Costa Chica's first Black Villages and Peoples Conference, called Encuentro de Pueblos Negros, in 1997. This description provides an important prelude to a discussion about the constitutional changes to the Mexican census.

## From Indigenous Autonomy to Black Recognition

The first alliance between Afromexicanos and Indigenous groups in the Costa Chica occurred within the context of agrarian struggles in the state of Guerrero. Black and Indigenous campesinos formed an organization in the Costa Chica de Guerrero called Consejo *Guerrerense 500 Años de Resistencia Indígena, Negra y Popular* (Gurrero Council 500 Years of Black and Indigenous Popular Resistance) in 1992 that coincided with 500 years of Black and European presence across the Americas. Just a year later, the Jamiltepec Center for Indigenous Development began broadcasting for several hours a week a radio program dedicated to negritud, or Black life, across the Costa Chica. The program still exists along with a second Black radio station in Cuajinicuilapa, Guerrero, that began broadcasting in 2002.

The rise in Black consciousness and the mobilization of Black Politics of Identity coalesced in 1997, fueled by local and global factors that synergized to bring together the first meeting of Black peoples and villages and the formation of important Black organizations. The creation of local NGOs, such as the Organizazión Política Afrodescendiente and México Negro AC (*http://facebook.com/MéxicoNegroAcMéxicoNegroAc*), in combination with the first *Encuentro de Pueblos Negros* (Reunion of Black Peoples and Villages) in El Ciruelo, Oaxaca, brought people from multiple walks of life from across the Costa Chica together for the first time. In its now annual meeting, the *Encuentro de Pueblos Negros* convenes African descendants representing agricultural and fishing *ejidos* and cattle cooperatives from across Black villages and hamlets of the region together with professionalized Afromexicanos from local NGOs and state officials. The meetings include representatives from the *Museo de las Culturas Afro-Mestizas* in Cuajinicuilapa and the *María Luisa Aguirre Palancares* cultural center in the city of Pinotepa Nacional. These are important regional cultural institutions that function to disseminate knowledge related to Black identity and the production of the politics of identity that inform the reconfiguration of regional identity in relation to Afromexicano cultures in the Costa Chica.

The important milestones discussed earlier occurred just 3 years after the Zapatista Revolt in the state of Chiapas, and constitutional changes allowing for the creation of indigenously governed municipalities were enacted. As the annual meetings continued and more local, state, national, and international groups participated, the coalitions compiled a set of objectives to guide their mission (Weltman-Cisneros, 2013), which includes the following two demands:

1. To discuss, review, and approve activities that allow for Afromexicanos to meet, to make themselves visible to Mexican society, and be acknowledged by the three levels of the government, with the purpose of fighting and ending the discrimination against various groups and individuals on a local, regional, and national level.
2. To seek alternatives for community and regional development in terms of investment, growth, resource development, and community and regional organization.

As a result of the efforts of *México Negro* and other groups, in 2011 the Ministry of Indigenous Affairs announced that it would create a Department of Afro-Mexican Community Affairs in Oaxaca. The legal recognition of these communities is a crucial step in achieving not only state, but more important, federal recognition of Afromexicanos. The creation of this agency is a testament to the willingness of Afromexicanos and Indigenous groups in Oaxaca to not only learn from one another but to collaborate as well.

At the annual *Encuentro de Pueblos Negros* meetings NGOs demanded that Afromexicanos be recognized by the state, be counted by the census, and be lobbied for their objectives. The Encuentro has taken place outside of the Costa Chica in Veracruz, in Mexico City, and in the northern state of Coahuila. The *Costeños*–Afromexicanos have acted in solidarity with local NGOs and have begun to form cross-scalar alliances at the local, regional, state, national, and even international level.

The work of these civil society groups came to fruition in 2015 when the federal government announced it would include the category Afromexicano on the national 2020 census. In 2015 the *Instituto Nacional de Estadística y Geografía* included a question about Blacks on its national mid-census survey. The survey asks, "based on your culture, history, and traditions, do you consider yourself Black, meaning Afromexicano or Afro-descendant?" (INEGI, 2019) This is a milestone for Afromexicano recognition. Blacks had not been counted or surveyed by the federal government since the 1810 census! By not including a question about identity or recognizing the category, Blacks were in effect erased from the national identity. This *invisibilización*, to use the Spanish term, of Afromexicanos removed any official recognition of Blacks or the responsibility for providing services and resources for this important segment of the population.

Finally, in 2019, the senate voted to constitutionally recognize Afromexicanos. This event brought much jubilation from Afromexicano organizer Gina Diédhiou. "Four centuries later, we are finally recognizing that we are here, we were born here and we've always been here . . . we were in tears because there was something special about seeing 122 people vote yes, not one person said no" (Bailey, 2019).

The following section describes debates about nomenclature among Costa Chica residents, specifically, what will they call themselves, and how the decision will dictate how communities and people from outside of the Costa Chica will refer to them.

## Fissures between Place and Identity

At the first meeting of *Pueblos Negros* the main topic centered on how Black *Costeños* would self-identify—an important discursive power tactic in relation to self-identification, nomenclature, and racial terminology—and equally important, how the *Costeños* wanted to be identified by others. Locally, many of the villagers fishing and tilling their crops across the region identified simply as *Costeños*. Simultaneously, *Costeños* took no offense to the terms *moreno* or *pardo*, which are used in Mexico to soften the descriptive *negro*, in reference to their appearance and identity (Jerry, 2013). As Vaughn (2005) describes in his ethnographic work he has conducted in Mexico over the last three decades, many Black *Costeños* in the 1980s and early 1990s attempted to distance themselves from Blackness by using the terms *pardo* and *moreno* instead of *negro*. While at the same time, *Costeños* racially and culturally differentiated themselves from the Mixteca Indigenous communities living in the foothills of the Sierra Madre Occidental and the elite mestizos and Europeans residing in the regions' two main cities by their rhythmic use of language and music (Ramsay, 2004), distinct customs, and their appearance (Vaughn, 2013).

With the rise in Black consciousness by professionalized *Costeños*–Afromexicanos, combined with an increase in Black awareness across Latin America and the first *Pueblos Negros* meeting, the term *negro* became more celebrated among NGO members and workshop organizers, who aimed to instill in *Costeños* a sense of negritud, or pride in Blackness. Indeed, the enthusiasm is palpable in the words of a female participant in a workshop on negritud, or Black life, " I think we need to celebrate our history and culture as negros. . . . it's a beautiful thing to learn about ourselves, to know that there is something good about being negro" (Vaughn, 2013).

This statement does not take into consideration the heterogeneity of Blackness and identity in the Costa Chica region of Mexico or across Latin America. If *Costeños* identify strictly as Afromexicano, does this signify they cannot mobilize intersectional identities and identify as campesino, or mestizo, or even Indigenous? The current reality is that there are fissures or discrepancies that exist in relation to the racial identification of the *Costeños* who labor in the mango plantations and who work their *milpas* or gather mollusks and fish, and the professionalized Afromexicanos who reside in Cuajinicuilapa or Pinotepa Nacional or who have made their home in the Costa Chica but are part of Black power movements across Latin America (personal communication). The situation raises the question of who is making the decisions for how people of African descent are to self-identify and be identified by other Mexicans, the Mexican state, and other countries. Other studies argue that the implementation of the term *negro* to identify peoples, organizations, and events—such as the NGO, *México Negro*, or the name for the annual meeting of people of African descent, *Encuentro de Pueblos Negros*—does not coincide with the multiple terms that people from the Costa Chica may use to identify themselves (Lewis, 2000).

I believe that the NGO activists and organizers reinvent the past in associating *Costeños* with the great African kings of West Africa and Egypt. True, it is important to link African diasporic communities across the Americas with the African continent. These same activists must also promote the *Costeños* contribution to the social landscapes of the Costa Chica and African descendants' contributions to the modern Mexican state. These achievements must be highlighted for Afromexicanos to improve their social and economic standing as Mexicans.

Each society socially produces race distinctly. The way that U.S. institutions have historically interpreted Blackness has left no room for ambiguity, although now even U.S. categories are changing, because there now exists the option to choose Black and Latino on the census. In Brazil, racial identity is fluid, and Brazilians often perform their racially distinct identities in different situations according to how it benefits them to do so. The 2020 Mexican Census includes a sentence in which Mexicans can voluntarily identify as Afromexicano. Nevertheless, socially—within 21st-century Mexican society—the adoption of U.S. standards of race essentializes the *Costeños* identities and does not allow for a coming together of the identities of race and place in this locale. It is important to be *pardo*, *moreno*, or *Costeño*, and to also have solidarity with African descendants from Veracruz and Coahuila at a national scale, or with Cuba, Columbia, Brazil, and even the United States at an international scale under the umbrella of Afromexicano. The Afro, in Afromexicano, links the *Costeños* with the African diasporic communities from across the Americas. But place and geography matter too. This discrepancy in nomenclature related to place—*Costeño*—and race—Afromexicano—will continue to unfold as the movement matures, youth become more educated and mobilized in relation to racial identities, and as professionalized Afromexicanos work with communities across the Costa Chica in relation to social and economic opportunity.

## Does Getting Counted Count?: Economic and Social Development in the Costa Chica

The 2020 census provides space for Afromexicanos to claim their Afromexicano identity. March 2020 coincided with launching the census across Mexico, and for the first time Mexico celebrated the month of March as Black History Month. Civil society has mobilized to educate people through social media and face-to-face workshops about the importance of representation in the census—how now race does not represent (officially on the census) exclusion. How, in 2020, self-identifying as African descendants will allow Afromexicano inclusion "in sectors of society to obtain equal rights and access to political representation, economic development and future progress." (Mexico Negro A.C., n.d.) The NGOs have worked with the national government to acquire funding to educate Mexicans on the possibility of self-identifying as Afromexicano. The Afro-Censo's-#AfroCensoMX (Black census) motto, *por su puesto que si* (but of course) to the question of self-identification as an Afromexicano encourages people to participate. The national government's campaign office, called AfroCenso MX, sponsored a song to promote the 2020 census changes, called *AfroMéxico Sí*! Susana Harp, a musician, performer, and now a senator, who leads a women's rights NGO in Oaxaca City and has contributed money to help create Afromexicano NGOs in the Costa Chica, wrote the lyrics, which includes phrases such as "it's not just a question of the color of your skin, we celebrate our contributions and heritage in Mexico." She did not sing it (perhaps because she is of elite Lebanese descent and a relative of Carlos Slim?). Alejandra Robles, an artist from Veracruz with noticeable African features, performs the song. The video provides images of *afrodescendentes* from across Mexico but mostly from Veracruz.

Will this act of being counted matter for *Costeños*–Afromexicanos in terms of geographies of development, socioeconomic transformation, and the creation of an officially recognized region?

Mexican technocrats have applied neoliberal political and economic development models for organizing Mexican society, the state, and economic activities. Following these neoliberal policies has catapulted Mexico's export sector, especially in the manufacturing of cars and electronics, from the maquiladoras at the northern border, but also in the exporting of seasonal fruit to U.S. consumers. While Mexico's GDP has increased and the average Mexican salary has gone up, these policies have led to more entrenched cleavages within Mexican society. The uneven geographies of development can be found at multiple levels within Mexican society but are especially glaring between the industrialized north and Mexico City compared with the south of the country, where the majority of Indigenous and Afromexicanos reside.

Neoliberal development strategies have strengthened since the implementation of the constitutional changes that recognized indigeneity and Indigenous autonomy and also the first *Pueblos Negros* meeting in 1997. Coinciding with the neoliberalization of Mexico, NGOs have become more prominent and neoliberalized (Walker et al., 2007). Alternative grassroots development has functioned within the informal and formal sectors of economic exchanges; thus neoliberal doctrine also impacts the spaces they move through—the lives, space, geographies—of Afromexicano NGOs in the Costa Chica.

The language used by *Mexico Negro* A.C., the grassroots NGO, concerning political representation, economic development, and future progress replicates the language of the neoliberal state and other NGOs. The question is what will this progress and development look like, and who will benefit. Will this take the form of grassroots development in the fishing and aquaculture industries and locally owned and operated tourist centers? Or will the development projects be overseen by Japanese fishing conglomerates and Canadian and Spanish hotels and resorts where the Afromexicanos living in Corralero and across the Costa Chica merely follow in the footsteps of their compadres and comrades, laboring in poor conditions across the thousands of maquiladoras located along the U. S.–Mexico border or in Spanish hotel chains built on former fishing *ejidos* like those found in Huatulco, watching their ecosystems and cultural landscapes deteriorate around them? Analyzing how Indigenous communities have improved since 1994 may provide an indicator of the successes of Afromexicano communities.

The 2019 constitutional changes that recognize Afromexicanos as part of a multicultural nation are a huge step. After 300 years of slavery and 200 years of being erased from the imagined national identity, Mexico's constitutional recognition of Afromexicanos is a feat! Afromexicanos have achieved much recognition in the 20-plus years of organizing for institutional recognition, and although "this coalition of intellectuals and grassroots activist still has [far] to go in addressing its history of racial discrimination, it also [provides] hope that this combination of successes—especially the new census question—sets us on a path toward significant advances." (Frias, 2020).

## DISCUSSION QUESTIONS

1 According to information in this chapter, Africans in Mexico (Afromexicanos) have been represented by the authorities and/or state distinctly throughout history. Describe how Africans and Afromestizos were categorized during the colonial period. Then explain what role Afromexicanos played in the war for independence in 1824. After gaining independence from Spain, how were Africans represented by the federalist state (from 1824–1910)? Society went through major changes after the Mexican Revolution (1910). The philosophy of *la raza cósmica* was intended to strengthen Mexican identity and solidify Mexican nationalism after the revolution and into the 20th century. What were the tenets of the *la raza cósmica*? Were Afromexicanos represented in the philosophy's blended identity? When did Afromexicanos finally gain full representation by the state in Mexico?

2 Can you provide any examples of shifting or fluid identities in your community or society (perhaps think about changes in the U.S. Census or the use of preferred pronouns)? Now describe (1) in what culture (s) do you participate; (2) to which racial or ethnic group (s) do you identify with; (3) what is the basis of your primary cultural identity; and (4) do others readily recognize this basis and identity, and why or why not?

3 According to the Vignette on Environmental Determinism, why do we still see the majority of contemporary African Diasporic populations centered in the Caribbean and the Caribbean Basin, the southeast of the United States, and in Northeastern and Southeastern Brazil, and how is this related to Environmental Determinism?

4 Explain how and why the divergence of cultures that we experience around the globe today is not determined solely by a society's physical geographical surroundings.

5 Describe the different stories of how people of African descent came to the Costa Chica. Now explain how the shipwrecked origin narrative unites the Black population of the Costa Chica.

6 Explain how and why Africans in west Africa, Indigenous groups farming the Chinampas, and Europeans chose which crops to grow in the tropics, why they chose these particular crops, and how these examples reveal how different cultural societies (West Africans, Amerindians, and Europeans) value and exploit natural resources distinctively in similar environmental conditions.

7 Explain the difference between identity politics and politics of identity.

8 Does the adoption of the term Afromexicano allow for an intersectional understanding of race and identity in the Costa Chica? Explain what intersectional identities may be of importance to Costa Chica residents that are not encapsulated under the term Afromexicano.

## SUGGESTED RESOURCES

To learn more about the African Diasporas across Latin America, see:

The Afro-Latin American Research Institute at Harvard University, *https://alari.fas.harvard.edu.*

For a comprehensive study of Afro-Mexico, see:

Vinson, B., and Restall, M. (2010). Black Mexico: Race and society from colonial to modern times. *African Diaspora Archaeology Newsletter, 13*(2), Article 35. Available at *https://scholarworks.umass.edu/adan/vol13/iss2/35.*

Director, Nicolas Segovia's documentary, *Costa Chica*, highlights the dances, music, and myths of the Afromexicanos of the Costa Chica. Available at *https://vimeo.com/171218439.*

*La Negrada*, or Black Mexicans, is the first Mexican fiction movie shot entirely with an all-Black Mexican cast throughout the towns and villages of the Costa Chica. The film provides an excellent view of the landscapes and livelihood strategies of the Afromexicanos of the Costa Chica. Available at *http://africanfilm.com/products/la-negrada-Black-mexicans.*

Archibold, R. C. (2014, October 25). Negro? Prieto? Moreno? A question of identity for Black Mexicans. *The New York Times.* Retrieved from *www.nytimes.com/2014/10/26/world/americas/negro-prieto-moreno-a-question-of-identity-for-black-mexicans.html?searchResultPosition=1.*

Dixon, K., and Burdick, J. (Eds.). (2012). *Comparative perspectives on Afro-Latin America.* Gainesville, FL: University of Florida Press.

Encyclopedia Britannica. (2017). Slave rebellions. *Encyclopedia Britannica.* Retrieved from *http://britannica.com/topic/slave-rebellions#ref1218356.*

Simms, E. Y. (2008). Miscegenation and racism: Afromexicans in colonial New Spain. *The Journal of Pan African Studies, 2*(3), 228.

Varagur, K. (2016). Mexico finally recognized its Black citizens, but that's just the beginning. Latino Voices, *Huffington Post.* Retrieved from *http://huffingtonpost.co.uk/entry/mexico-finally-recognized-its-black-citizens-but-thats-just-the-beginning_n_568d2d9ce4b0c8beacf50f6b.*

Vaughn, B., and Vinson, B. (2011). Unfinished migrations: From the Mexican South to the American South—Impressions on Afromexican migration to North Carolina. In D. J. Davies (Ed.), *Beyond slavery: The multilayered legacy of Africans in Latin American and the Caribbean* (pp. 223–245). Lanham, MD: Rowman and Littlefield.

Velázquez, M. E. (2024). *Africans and Afro-descendants in Mexico and Central America: Overview and challenges of studies of their past and present.* Collective Volume The Slave Route Project, UNESCO.

Vulloch, A. (2009). *Afromexicans: A short study on identity.* Unpublished manuscript, Graduate degree program in anthropology and the graduate faculty of the University of Kansas.

## REFERENCES

Aguilar Rangel, J. (2022). Infographic: Afrodescendants in Mexico. *www.wilsoncenter.org/article/infographic-afrodescendants-mexico.* Last accessed 07/07/2024.

Aguirre Beltrán, G. (1972). *La población negra de México: Estudio etnohistórico*. Fonda de cultural económica: Mexico City: Fondo de Cultura Económica.

Bailey, E. (2019). Afro-Mexicans may finally get recognition in the Mexican constitution, but many say that's not enough. Retrieved from *https://remezcla.com/features/culture/Afromexicans-constitution-recognition*.

Chao Romero, R. (2010). *The Chinese in Mexico, 1882–1940*. Tucson: University of Arizona Press.

Curry, A., and Bénali, R. (2019, July 14). The first Europeans weren't who you might think. *National Geographic*. Retrieved from *http://nationalgeographic.com/culture/article/first-europeans-immigrants-genetic-testing-feature*.

Diariomo. (2018, October 30). *Susana Harp y el Arrojo para Cruzar la Línea Entre Arte y Política—Diario* Moneda. Author.

Esteva, G. (2005). Celebration of Zapatismo. *Humboldt Journal of Social Relations, 29*(1), 127–167.

Frias, S. (2020, March 20). *Mexicans of African descent are being counted in historic 2020 census*. Every Child Thrives. Available at *https://everychildthrives.com/y-ap-konte-meksiken-ki-dorijin-afriken-yo-nan-resansman-2020-istorik-la/?lang=ht*.

General Archive of the Indies. (2022). General Archive of the Indies (Mexico). Retrieved January 9, 2019, from *https://en.andalucia.org/listing/general-archive-of-the-indies/15741101*.

Gersmehl, P. (2014). *Teaching geography* (3rd ed.). New York: Guilford Press.

Hall, S., and Du Gay, P. (1996). *Questions of cultural identity*. London: Sage.

Hooker, J. (2005). Indigenous inclusion/Black exclusion: Race, ethnicity and multicultural citizenship in Latin America. *Journal of Latin American Studies, 37*, 285–310.

Inikori, J. E. (2020). Atlantic slavery and the rise of the capitalist global economy. *Current Anthropology, 61*(S22).

Instituto Nacional de Estadística y Geografía (INEGI). (2019). Retrieved from *www.inegi.org.mx*, accessed 07/08/2019.

Jerry, A. R. (2013). Talking about Mestizaje: History, value, and the racial present. *The Journal of Pan African Studies*, 6(1),110–124.

Lewis, L. A. (2000). Blacks, Black Indians, Afromexicans: The dynamics of race, nation, and identity in a Mexican Moreno community (Guerrero). *American Ethnologist, 27*(4), 898–926.

Lewis, L. A. (2016). Indian allies and white antagonists: Toward an alternative mestizaje on Mexico's Costa Chica. *Latin American and Caribbean Ethnic Studies, 11*(3), 222–241.

Machado, M. H. P. T., dos Santos Gomes, F., and Ariza, M. B. A. (2019). *Quilombos*. New York: Oxford University Press.

Mann, C. (2005). *1491, Second Edition: New revelations of the Americas before Columbus*. New York: Vintage Books.

Mansferrer Leon, C. (2018, February 28). Afrodescendientes en los libros de texto de la CDMX. *Chilango*. Retrieved from *http://chilango.com/opinion/afrodescendientes-en-los-libros-de-texto/?fbclid=IwAR2YQLbp2dKi7OIN5mTZZEsdcdAoL9oqx-qwPk3RNViQyOoKIPm6l9DC5KE*.

Mexico Negro A. C. (n.d.). Available at *http://facebook.com/MexicoNegroAc/?locale=es_LA*.

McKittrick, K. (2006). *Demonic grounds: Black women and the cartographies of struggle*. Minneapolis: University of Minnesota Press.

Mignolo, W. (2009). Epistemic disobedience, independent thought, and decolonial freedom. *Theory, Culture and Society*, *26*(7–8), 1–23.

Muñoz, A. A. (2004). Explaining the politics of recognition of ethnic diversity and Indigenous Peoples' rights in Oaxaca, Mexico. *Bulletin of Latin American Research*, *23*(4), 414–433

Oaxaca State Government. (2020). Zoques–etnografías. *Atlas de los Pueblos Indígenas de México*. Mexico City: INPI. Retrieved from *https://atlas.inpi.gob.mx/zoques-etnografias*.

Oaxaca State Legislature LVI (Legislatura del Estado LVI). (1998a). Derechos Indígenas en la legislacio´n oaxaqueña [Indigenous rights in the Oaxacan law]. *Cuadernos de la Comisión Permanente de Asuntos Indígenas* [Volumes of the Permanent Commission of Indigenous Affairs], no. 3.FFF. Author.

Oaxaca State Legislature LVI (Legislatura del Estado LVI). (1998b). Iniciativa de ley de derechos de los pueblos y comunidades indígenas del Estado de Oaxaca [An initiative for the legal rights of indigenous towns and communities of the State of Oaxaca]. *Cuadernos de la Comisión Permanente de Asuntos Indígenas* [Volumes of the Permanent Commission of Indigenous Affairs], no. 4. Author.

Omi, M., and Winant, H. (1994). *Racial formation in the United States: From the 1960s to the 1990s* (2nd ed). New York: Routledge.

Ramsay, P. A. (2004). History, violence and self-glorification in Afro-Mexican "corridos" from Costa Chica de Guerrero. *Bulletin of Latin American Research*, *23*(4), 446–464.

Rodríguez López, A. (n.d.). "El Naufragio." Aydee Rodriguez López en pintoresmexicanos.com. Retrieved June 19, 2024, from *http://pintoresmexicanos.com/aydeerodriguez*.

Rodríguez-Pose, A., and Gill, N. (2004). Is there a global link between regional disparities and devolution? *Environment and Planning* A, *36*, 2097–2117.

Stewart, A. (2013). La chilena mexicana es peruana: Multiculturalism, regionalism, and transnational musical currents in the Hispanic Pacific. *Latin American Music Review/Revista de Música Latinoamericana*, *34*(1).

Vasconcelos, J. (1997). *The cosmic race/La raza cósmica (Race in the Americas)*. Baltimore: Johns Hopkins University Press.

Vaughn, B. (2005). Afro-Mexico: Blacks, Indígenas, politics, and the greater diaspora. In A. Dzidzienyo and S. Oboler (Eds.), *Neither enemies nor friends: Latinos, Blacks, Afro-Latinos* (pp. 117–136). New York: Palgrave Macmillan.

Vaughn, B. (2013). Mexico negro: From the shadows of nationalist Mestizaje to new possibilities in Afromexican identity. *The Journal of Pan African Studies*, 6(1).

Vinson, B., and Vaughn, B. (2004). *Afroméxico: El pulso de la población negra en México: Una historia recordada, olvidada y vuelta a recordar.* Mexico City: Fondo de Cultura Económica.

Walker, D., Jones, J. P., Roberts, S. M., and Fröhling, O., R. (2007). When participation meets empowerment: The WWF and the politics of invitation in the Chimalapas, Mexico. *Annals of the Association of American Geographers*, 97(2), 423–444.

Weltman-Cisneros, T. (2013). Cimarronaje cultural: Towards a counter-cartography of Blackness and belonging in Mexico. *The Journal of Pan African Studies*, 6(1), 125–139.

CHAPTER 4

# The Social Geographies of Tijuana

## *Urbanization at the U.S.–Mexico Border*

### Migration, Irregular Settlements, and Neighborhood Formation

The mountains of Sinaloa form part of the Sierra Madre Oriental that separate it from the eastern neighboring states of Chihuahua and Durango. As the Sierra Madre descends toward the west, beyond the foothills lays a narrow coastal plain. The state's economy has historically consisted of fishing, agriculture, brewing beer, and smuggling. Native Sinaloense Jesus Malverde, the 19th-century equivalent of a Sinaloa Robin Hood, has been proclaimed the patron saint of smuggling by the men and women who choose this profession. There is a shrine honoring him in the center of Culiacan, the capital city of Sinaloa.

The history of smuggling can be traced to illegally importing weapons south from the United States during two periods of French intervention and during the long Mexican Revolution (1910–1936) to distribute to Mexican armies across the republic. The foothills of the Sierra Madre resemble the opium poppy-growing regions of Afghanistan and Burma. When the United States had no access to opioids at the peak of World War II, the Army Corps of Engineers planted poppies in the mountains of Sinaloa in order to access morphine for conducting operations at domestic hospitals and at the war fronts. After the war, the poppies continued to bloom, and soldiers returned to the United States addicted to opioids, so the Sinaloense smugglers began bringing heroin and marijuana north across the border to feed the U.S. drug appetite. Sinaloa is home to many of the past and current drug cartel leaders who learned the smuggling trade from their fathers and uncles, who in turn learned it from their forefathers.

This is the scenario that Adelina's father, Chuy, faced as he pondered his options and made the decision to move to Tijuana. People don't generally migrate when things are going well. There are push-and-pull factors that would-be migrants consider when making the important decision to move. The push factors listed earlier, including having been born into a region fraught with little opportunity aside from working in tourism in coastal areas, laboring in breweries, or joining the smugglers, gave Adelina's father reason to leave, and Tijuana's burgeoning economy during the 1950s provided the pull factor.

Chuy settled with his wife and two sons in Cartolandia, the irregular settlement previously located in the River Zone. From there he participated with an organized group of squatters in an *invasion pirata* (essentially organized squatting) on empty land that forms an arroyo through a canyon that rises up west from the River Zone above the well-plotted downtown and Teniente Guerrero Park. He built a house in situ and continued to add on to it as it morphed into a comfortable family home where his daughter, Adelina, was born. As the neighborhood matured and regularized, the residents dubbed their community *Independencia*, a place name that openly states the entrepreneurial and independent spirit of the community's founders. Chuy named the street where Adelina grew up, Calle del Pino, in honor of the lone pine tree planted on the street.

Adelina inherited the large home with land-title deeds, electricity, and running water. She and her husband Ricardo—who commutes across the border to work as a gardener at San Diego State University—raised their six children there. All her children attended universities both in Mexico and the United States and are professionals working on both sides of *la linea*, or the border, as it is colloquially referred to in Tijuana. As their family grew and they required additional income, Adelina and Ricardo built a six-unit apartment building in the backyard of their home adjacent to the 15-foot plantain bush that her father had planted there years before, perhaps to remind himself of forays to coastal communities of the Sinaloa of his youth. The second-floor apartments face northwest, with views of the Tijuana estuary and, on clear days, the downtown San Diego skyline.

Chuy left the foothills of Sinaloa to avoid what he perceived may have been a life negotiating with narcotrafficking, the ramifications of which often include narcoviolence and perhaps rural poverty, or working at the Modelo Brewery in Culiacan or Mazatlán, because he had a dream that his life, and the life of his children, could be something more. Chuy, Adelina, her husband Ricardo, and their children's lives represent one story out of a million stories that embody the contemporary social geographies of Tijuana. In Tijuana people are capable of producing spaces of dignity where their independence can be achieved in *la colonia independencia.*

## Introduction

This chapter explores how Tijuana's social, political, and physical geography have shaped the economies, cultures, and urban morphology of the city, as well as how the spaces of the city impact its residents and the Tijuana–San Diego border region—geography matters. The chapter begins by identifying Tijuana's key location and the environmental consequences of the city's topography. It then discusses how the city has been shaped by policy decisions made at the national level and in the United States, coupled with domestic decisions made by the U.S. government that shaped social–economic life across the United States and San Diego, impacting the social geographies of Tijuana. The role of capitalists and Mexican migrants in vice tourism is then explored, followed by a discussion of the social geographies of the irregular settlements, the National Frontier Program (PRONAF), and the border industrialization plan (BIP), all of which have contributed to the distinctive urban morphology of the city. Tijuana's rapid inward migration and population growth helped fuel the city's socioeconomic diversity that is produced by *Tijuanenses,* as exemplified in their music and gastronomy, and the emergence of a creative class that is reshaping the built environment of the city. Finally, the chapter considers how Tijuana, although still fraught with environmental degradation and socioeconomical cleavages, to many has become the Mexican dream.

## Where Is Tijuana?

Tijuana is a city of 3 million people that sits at the furthest northwestern corner of Mexico. Toward the west of the city is the Pacific Ocean, a great waterway that separates and links communities across the Pacific Rim and beyond (see Figure 4.1). The town of Tecate, best known for its namesake beer, is the first settlement found to the east of Tijuana. Southeast of Tijuana, separated by coastal mountains from the sea, are the wineries and vineyards of the Guadalupe Valley, where Spanish friars planted the first vines at Mission Santo Tomas Aquino in 1791. The port city of Ensenada is located on the Bay of Todos los Santos south of the Guadalupe Valley. The modernized port concentrates on fishing, canning, and container shipping. Cargo ships arrive from North America, Latin America, and Asia with components and parts that are shipped by road and train to Mexicali, Tecate, and Tijuana to be assembled by

FIGURE 4.1 The fence designating the international boundary disappearing into the Pacific Ocean. Photo by David M. Walker.

predominantly female workers in the maquiladoras and reexported to consumers across the Global North. Toward the north is California, which became a state in 1849 after the "war of the Northern invasion," as it is known in Mexico, the signing of the Treaty of Guadalupe Hidalgo, and the Gadsden Purchase established the border between the United States and Mexico.

A bustling border city in Mexico that faces the economically and culturally dynamic city and county of San Diego, Tijuana has three border crossings with a fourth planned to open in 2026 (San Diego Association of Governments; SANDAG, 2023). The San Ysidro/Puerta Mexico Crossing provides access for pedestrians, cars, and bicycles. Mexican entrepreneurs rent the bikes to border crossers so they do not have to wait in the long pedestrian lines. The entrepreneur accompanies the border commuter across the international boundary, also on a bicycle. Once the border commuter makes the trip across the border, the bike renter returns into Mexico to provide the service to the next border crosser. The Otay Mesa/Mesa de Otay crossing accommodates the majority of freight travel between the industrial borough of Otay in Tijuana with its namesake municipality on the California side of the border. The Otay Mesa East II border crossing slated to open in 2026 will help ease the freight congestion at the border. Unlike the other ports of entry, such as San Ysidro, a toll will be required when exiting the inspection booth on the U.S. side. The money from the toll will fund infrastructure projects on both sides of the border, and according to authorities, the toll will guarantee an average waiting time of 20 minutes. The fourth crossing, the Cross Border Xpress, is a 390-ft.-long bridge that connects the Tijuana International Airport, which flanks the international line, with a service terminal in San Diego. (International border crossings, and the airport location, are shown in Figure 4.2.) This service is not free, and travelers must be using the Tijuana airport to fly to destinations across Mexico. The cost is roughly $40 for a round-trip crossing in either direction.

The binational San Diego–Tijuana region consists of 40 percent of the entire United States and Mexico border population and the busiest land border crossings in the western hemisphere (if not the world), with approximately 120,000 passenger vehicles, 63,000 pedestrians, and 6,000 trucks crossing back and forth every day. This transborder megalopolis is a $230 billion economic engine with over 6 million residents and nearly 2 million employees.

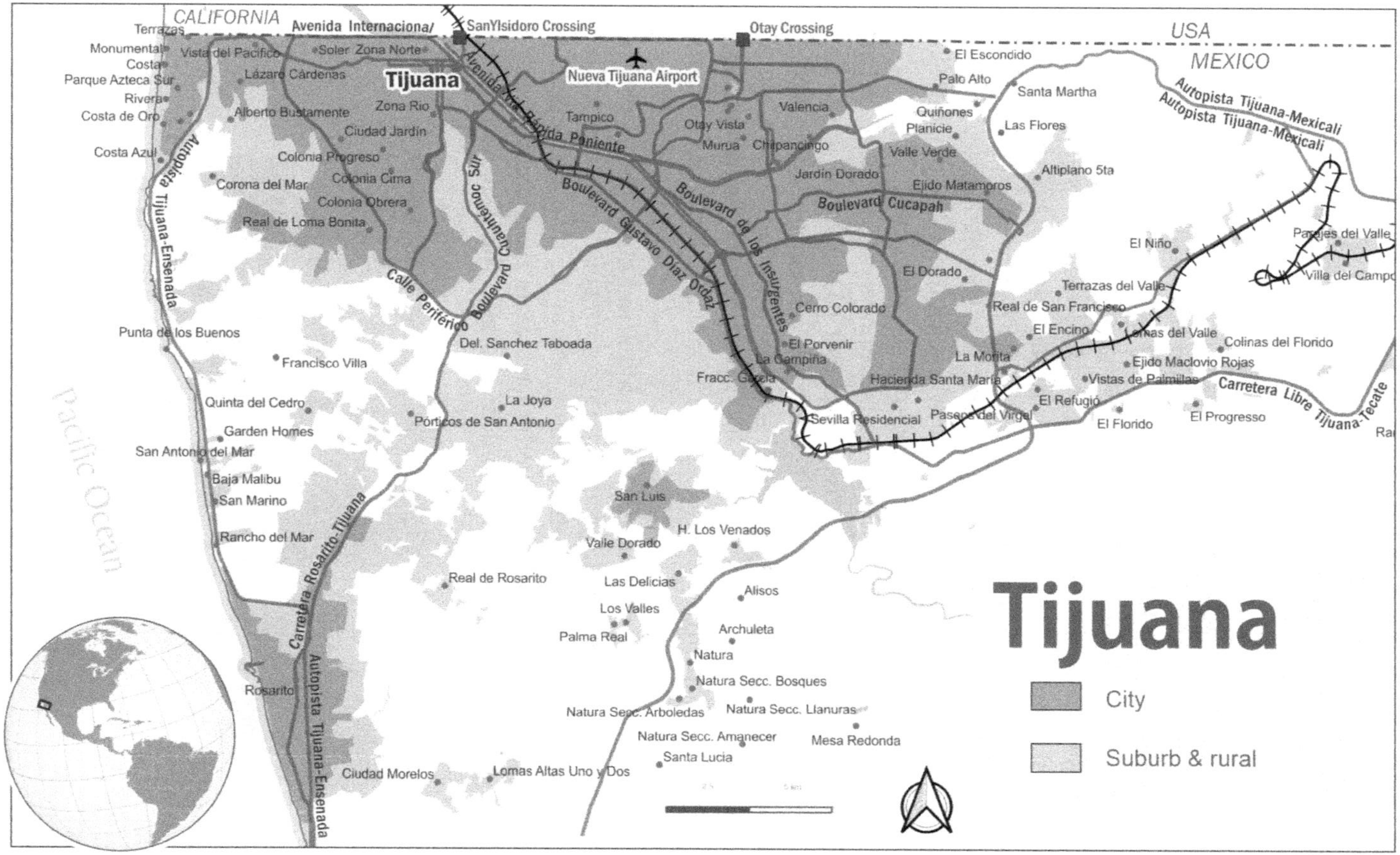


FIGURE 4.2 Map depicting the greater Tijuana metropolis with international border crossings and the international airport.

## Tijuana's Topography and Its Impacts on the Environment

The topography of the Tijuana–San Diego border impacts the built environment and ecology of this binational region. The confluence of the Arroyo de Palomar and Arroyo Las Palmas rivers, which form the headwaters of the Tijuana River in the Sierra de Juarez in California on the U.S. side of the border, crosses south into Baja California, Mexico. The Tijuana Rver passes through rugged and challenging topographies of canyons, mountains, ravines, and mesas of the municipalities of Tecate and Tijuana, before it flows northwest near San Ysidro, California, where it crosses the border a second time to form the Tijuana Estuary in San Diego County. Hence, San Diego County in California and the municipalities of Tecate and Tijuana share a binational watershed.

Tijuana sits on a mesa and a coastal mountain range divided by the Tijuana River. The river flows from southeast to northwest across the international border through the Tijuana Estuary, emptying into Imperial Beach, the southernmost coastal settlement in San Diego County (see Figure 4.3). The city's topographical position at a higher altitude that faces the border has negative environmental and economic impacts on the region.

The rivers flow through the Otay Mesa industrial parks, replete with hundreds of maquiladoras and dozens of irregular settlements. The contamination from the industrial pollution form the maquiladoras, and the renegade sewage flowing from the latrines of the irregular settlements impacts both sides of the border, especially the Tijuana Estuary

FIGURE 4.3 The Tijuana River and Estuary at Imperial Beach, California. Tijuana is in the background. Photo by David M. Walker.

and Imperial Beach in San Diego County. This topographical juxtaposition at the border combined with Tijuana's industrialization and population growth have led to contaminants from the unregulated maquiladora industry and raw sewage from irregular settlements to flow across the international boundary into the Tijuana River Estuary located in the United States, leading to the closure of San Diego's beaches. Beach closures negatively impact tourism dollars and create economic hardships for commercial and sports fishing along the coast. This contamination also destroys the habitat at the Tijuana River Estuary, home to 29 species of fish, 29 species of reptiles and amphibians, and 370 species of birds. The estuary also protects 10 endangered species, including the San Diego fairy shrimp, the light-footed clapper rail, and the salt marsh bird's beak, and is an important stopover spot for migrating birds. Ironically, this topographical difference provides many of the irregular settlements with fantastic coastal views as far as Cabrillo Point and the downtown San Diego skyline and an ocean breeze to help clear the air of dust and pollution.

In contrast to the river's trajectory, dictated by topography and gravity, it takes little physical effort to draw the straight line that divides two countries on a map. In the case of the boundary between California and Mexico, this topographical reality prevents the easy enforcement of border crossings. Indeed, the border fence extends from Jacumba, California, 27 miles east of Playas de Tijuana, until it ends at the Pacific Ocean. The fence continues again further east of Jacumba near to the town of Tecate. It extends past Tecate further east for another 40 miles until the terrain becomes too rugged and prevents the construction of a border wall or a fence along the boundary. Along the California–Mexico border the fence already exists where it was physically feasible to build it. Border patrol agents from Mexicali to San Diego don't frankly see where the wall can be expanded within their sector. The topography does not permit it.

Tijuana is located topographically on a mesa and a coastal mountain range divided by the Tijuana River that flows from southeast to west–northwest across the international border through the Tijuana Estuary, emptying into Imperial Beach, the southernmost coastal settlement in San Diego County (see Figure 4.3). The city's topographical position at a higher altitude that faces the border has negative environmental and economic impacts on the region. This topographical juxtaposition at the border combined with Tijuana's industrialization and population growth has led to contaminants from the unregulated maquiladora industry and raw sewage from irregular settlements to flow across the international boundary into the Tijuana River Estuary located in the United States, leading to the closure of San Diego's beaches. Beach closures negatively impact tourism dollars and create economic hardships for commercial and sports fishing along the coast. This contamination also destroys the habitat at the Tijuana River Estuary, home to 29 species of fish, 29 species of reptiles and amphibians, and 370 species of birds. The estuary also protects 10 endangered species, including the San Diego fairy shrimp, the light-footed clapper rail, and the salt marsh bird's beak, and is an important stopover spot for migrating birds. Ironically, this topographical difference provides many of the irregular settlements with fantastic coastal views as far as Cabrillo Point and the downtown San Diego skyline and an ocean breeze to help clear the air of dust and pollution.

## A City Shaped from Both Sides

Since Tijuana caught the attention of U.S. citizens in 1898, at the end of the Spanish American War, and firmly got Americans hooked by the 1920s, journalists, politicians, and social commentators like to describe Tijuana as a chaotic city, replete with organized crime, rampant drinking, prostitution, violence, poverty associated with narcotics trafficking, and environmental degradation. Similarly, the central government in Mexico City has often viewed Tijuana as a city whose residents have been corrupted by their proximity to the United States and therefore aren't any longer "truly" Mexicans. Mexico's centralized political system and Tijuana's great distance from the center of power in Mexico City have contributed to the political and economic abandonment of this border city by the central government from its inception, until economic opportunity changed this perspective in the 1960s. In the 21st century these U.S. perceptions of Tijuana as only a vice-ridden city and of Mexico City viewing Tijuana as a North American-corrupted outpost have changed.

The Tijuana of the 21st century houses universities, financial centers, and a thriving arts and culture scene. The city offers fantastic gastronomy, from tacos on the street corner to family run *fondas* and cantinas to some of the most upscale restaurants found in Latin America. Tijuana's hinterland provides the city with wines from the Guadalupe Valley, an abundance of seafood from the fishing fleet and the fish market in Ensenada, lobster from Puerto Nuevo and Popotla, fresh vegetables and greens from San Quintin, and hormone-free chicken, duck, and quail from boutique farm-to-table producers scattered across Baja California Norte.

Yes, Tijuana is a major center for the assembly of electronics and other goods that are sold primarily in the Global North, which has played a role in shaping the economy, culture, urban morphology, and environmental degradation of the city. Concurrently, Tijuana's rapid diversification is producing a complex and distinctive urban society that is quintessentially Mexican. The city does have strong ties to San Diego, Los Angeles, and places beyond, but the daily reality of Tijuana is one that is produced by its residents, who are *Mexicanos*, in Mexico. And although over 65,000 people cross the border from Tijuana to work or go to school every day, and it is estimated that 40 percent can cross the border for shopping errands and social visits in southern California, most Tijuana residents never or rarely cross the border. Their quotidian lives exist entirely within Mexico, and their daily struggles shape the city's society and urban space in the 21st century even though they may live a stone's throw from the United States. Tijuana, indeed, is a very Mexican city.

Decisions made in Washington D. C. and Mexico City, combined with decisions made by residents on the ground, form the material dialectic that has produced Tijuana's character, cultural landscapes, urban structure, and built environment. Why did so many people move to this city in the 20th century and continue to do so in the 21st century? What are the external factors that correlated with this population growth? In order understand these questions, we need to consider how the demarcation of the international border between the United States and Mexico, and the growth of Tijuana, was shaped by war, U.S. foreign and domestic policies, and Mexican domestic policies at the border, combined with how *Tijuanenses* corroborated in the emergence and production of the complicated social geographies of this city.

## Vice Tourism and Urban Expansion

Tijuana's spatial structure, its built environment, and cultural landscapes changed in accordance with social and economic developments north of the border combined with the steady flow of labor arriving from the south. From the Treaty of Guadalupe Hidalgo to the end of World War II, migration from the interior of Mexico, U.S. domestic and foreign policies, and the dynamic capitalist economy of southern California fueled Tijuana's urbanization.

The Spanish American War in 1898, which opened the Pacific theater to U.S. imperialism with the establishment of military bases in San Diego County, marks the emergent role of Tijuana as an adult, vice-driven tourist destination produced by and for U.S. citizens. U.S. capital funded the construction of the first large-scale bars and brothels that were built to attract sailors stationed at the San Diego Naval Base. In 1906 San Diego newspaper and businessman John Spreckels built a railroad line from Yuma, Arizona, to San Diego. The line crossed the Mexican border and connected the border towns of Mexicali, Tecate, Tijuana, and San Diego. Two years later the Mexican government sanctioned gambling in Baja California, which spurred an increase in vice tourism from southern California.

In 1916 James Cofforth, a boxing promoter from San Francisco, constructed the first racetrack in Tijuana 400 yards from the international border. He not only paid for the construction of the new hippodrome, but he also paid for infrastructure improvements such as roads and the construction of a bridge across the Tijuana River. Three cabaret operators forced out of the Bakersfield area by a reform movement took their capital to Tijuana. Marvin Allen, Frank B. Beyer, and Carl Withington formed the ABW Corporation that would go on to fund cabarets, brothels, bars, and gambling across the city.

U.S. prohibition laws passed in 1919 ushered in the "golden era" of vice tourism in Tijuana. Breweries and wineries were constructed to quench the thirst of the growing clientele arriving from the dry counties of California and beyond. Hotels were built—such as the legendary Hotel Caesar on Revolution Avenue (the birthplace of the Caesar salad)—to accommodate the crowds. Famous Hollywood stars, Chicago gangsters, and sports figures visited the city, making Tijuana easily the most visited place in Mexico by U.S. nationals. At this time, Tijuana's uneven geographies of urbanization began to emerge as evidenced by the socioeconomic cleavages found across the city's landscape. Mexican nationals migrated from the interior of the country to work in the breweries and wineries, hotels, brothels, bars, cabarets, and gambling centers on Revolution Avenue—the area between the border and the racetrack—and at the Agua Caliente Casino in the newly established city. Mexican residents lived west of the tourism zone in the canyons and ravines jutting up from the tourist strip of Revolution Avenue and in the floodplain of the Tijuana River in shanties, without sewage infrastructure and running water.

Juxtaposed to the irregular settlements, the vice-tourism district of the city was an imitation of something that didn't exist: a combination of gringos' imagination of "Old Mexico" with flamboyant architectural designs and flourishes to create a sense of the exotic. These patterns of uneven development continue to replicate as the city diversifies its economic sector to maquiladora production and eventually to culture-based economies.

The end of prohibition in the United States lessened border towns' comparative advantage in luring thirsty thrill seekers from north of the border. In 1934, Mexico elected Lazaro Cardenas as President, who famously nationalized the oil industry in Mexico, but also tried to break Baja California's strong ties to the United States by ordering all gambling establishments and brothels closed in 1934. But by then, Tijuana had firmly established itself as a city infused with cultural and economic connections to communities in the south of Mexico, southern California, and beyond. When the casinos shut down, Mexicans searched for work *en el norte*. The money immigrants earned fueled the growth of neighborhoods flanking the border. During the 1950s and 1960s the irregular settlements of Libertad, Independencia, Obregon, and the Coahuila became important working-class neighborhoods that still exist today.

## Social Geographies of Housing in Tijuana: Irregular Settlements

Indeed, the majority of neighborhoods in Tijuana are irregular settlements that are created by the municipal planning department without zoning regulations or plans to install water, sewage and storm drains, and electrical infrastructure before the construction of homes begin. Demographic growth has been swift in Tijuana (see Figure 4.8, later in the chapter). The population has grown faster than the municipality's ability to provide housing and services. Investments in economic sectors of the city, from vice tourism to the maquiladora industries, have been made by foreign capital—people who often don't care how *Tijuanenses* live. Although direct foreign investment in Tijuana produces jobs, it regularly does not produce high tax revenues. Tax incentives and subsidies are used to entice business to the border—especially in the maquiladora sector. And because of government centralism in Mexico and the megagrowth of Mexico City (the phenomenon of urban primacy), which enjoys the best health care, subsidized transportation, and housing in the country, for every dollar paid in taxes to the federal government in Mexico City by the states and municipalities only 16 cents return to the provinces in infrastructure investment.

Mexican migrants to Tijuana have literally had to build their own homes and neighborhoods by hand (this is why irregular settlements are also called self-help housing in Mexico). The two principal models of irregular housing development in Mexico, including in Tijuana, are the invasion of land by squatters and the illegal subdivision and sale of lots by landowners. Land invasions can occur on a large scale by organized groups of squatters who settle a particular area in a coordinated fashion or on a lot-by-lot basis. In general, large-scale squatting, known in Mexico as *invasion pirata* (pirate invasion), occurred in Tijuana during periods of accelerated population growth. *Invasión pirata* squatters often have a political agenda, arrive with prepared demands, and are well organized in moving through the procedure toward regularization, or securing tenure. The form of squatting known as *invasión hormiga* (ant invasion), which is piecemeal land invasions, tends to occur in areas where residential development is not permitted by law due to steep slopes or other environmental conditions. The second mode of irregular development involves the illegal subdivision and sale of land without regard for planning regulations or for registering transactions.

To complicate the land-tenure issues surrounding these irregular settlements, often communal farmers presiding over *ejidos* illegally subdivide their communal farms. *Ejidos* are grants of lands presented to a group of individuals by the federal government. Once the state has granted the *ejido*, the *ejidatario* has property rights to his or her individual plot as well as access to lands held collectively by the *ejido* community. The farmer is allowed one urban plot—to build a house or run a store or school—and an individual rural plot of up to 25 hectares (about 61 acres). Communal farmers *(ejidatarios)* residing adjacent to rapidly growing cities often found it more lucrative to sell or rent their lands rather than work them for agricultural purposes. What's more, since the lands were illegally sold or rented, the *ejidatario* could claim that the lands had been invaded and demand double payment. If they were denied the money, they could complain to the state and seek legal counsel. This issue occurred frequently in Tijuana. More than half of the urban built environment of Tijuana sits on *ejido* property (Walker, 2000).

Upon arriving to Tijuana from the interior, Mexican migrants "invade properties" (effectively squatting on government-held private or communal land) or purchase a lot illegally and begin to construct their homes. At first, residents use any materials they can find, such as cardboard and sheet metal. Tijuana's proximity to southern California has spurred a brisk industry in recycled materials from north of the border. Many irregular communities use garage doors for their first homes. It is common to see homes built with four garage doors, a sheet-metal roof, and a dirt floor. The residents dig a latrine behind their homes. They purchase bottled water and butane gas containers for bathing, consumption, and cooking, although it is still common to see outdoor fire pits. Entrepreneurs drive through the settlements selling these items as well as many other goods (furniture, cooking wares, garage doors, etc.). The residents ingeniously (and dangerously) pirate electricity from existing infrastructure. That is how they begin their lives in these irregular settlements.

As the residents begin to earn and save money from working at jobs that provide higher incomes (moving from working as a shoe shiner to establishing your own taco cart) or use remittances sent by a relative working in the United States, they will begin to make improvements on their homes. Slowly the garage door walls will be replaced (and sold to a new builder) by concrete cinder blocks, the dirt floor will be covered with wood and perhaps Formica (that may also be recycled from *el norte*). Next plaster will be placed on the outer walls of the structure. As more children are born, a second story will be built. The houses continue to be improved until they have beautiful paint jobs, landscaped yards, and vegetable gardens. The homes are always in a process of becoming—and are never finished. This organic urbanization provides fantastically varied architectural styles in the irregular settlements (see Figure 4.4 for an example). According to each home-builder's fortune and style, you can have a house that looks like it was transported from a new development in southern California next to a home that is in the in-between construction space of half-garage door and half-cinder block walls. Again, recycled materials from southern California are used extensively. Many Sears and Roebuck Craftsman homes from the early 20th century were razed in the building of California's highway infrastructure. It is common to see a craftsman home as the second story of a house perched upon the first-story cinder-block structure.

How do the residents of irregular settlements in Tijuana "regularize" or get titles and services for their homes? Interestingly, land regularization in Mexico is often a politically driven process. Squatting is tolerated as a form of gathering support for the urban poor. In the case of Mexico, through a clientilistic system of favors for votes, the political party the

FIGURE 4.4 This picture shows an example of a home in an irregular settlement, or self-help housing, in Tijuana. The building on the right's ground floor has now been converted to a store, and the owner and her family have moved up to the second-floor home, brightly painted a vernacular yellow. Notice the tires initially brought to Tijuana from California to be recycled. Instead they have become a retaining wall to prevent erosion. Photo by David M. Walker.

PRI and the Partido Acción Nacional (PAN) party, which governed Tijuana from 1989–2004, 2007–2010, and 2016–2019 (Morena has governed Tijuana since 2019), have been able to continually "win" elections and govern at the national level for the large part of the 20th century (in the case of the PRI) and at the state and municipal levels in Baja California and Tijuana (in the case of the PAN). One favor that was traded for votes was access to urban land and services. Subsequently, regularization and titling programs were instituted in these neighborhoods for continuing political support.

Irregular settlements differ greatly from slums or shanty towns. People are proud of, respect, and love the homes they build in the irregular settlements of Tijuana. Sixty percent of the population lives in these communities.

The following section describes how in the second half of the 20th century the federal government in Mexico City began to invest in Tijuana. These initial investments would greatly change the landscapes of the city, attract additional migration from the south, and provide it with the Mexican cosmopolitanism that produces the social geographies of the 21st century.

## PRONAF and Geographies of Uneven Urbanization

The federal government in Mexico City began an ambitious project to modernize the points of entry along the U.S.–Mexico border called PRONAF, *Proyecto Nacional Fronterizo*. Established in 1961 by then-president Adolfo López Mateos, PRONAF sought to regularize commerce, trade, and industrialization of the border region. Its main goal was to strengthen economic and tourist activities of Mexico's northern border by implementing urban-infrastructure projects to modernize the city and provide industrial parks to attract transnational assembly plants or maquiladoras. PRONAF investments in the infrastructure for industrial parks, which helped to accelerate the growth of the maquiladora-assembly plants across Tijuana and the entire U.S.–Mexico border from 1961 until the codification of free trade with the signing of NAFTA in 1994 greatly impacted the built environment and social geographies of Tijuana. The growth of the maquiladora industry spurred inward migration from the south and center of the country to Tijuana. The rapid population growth and industrialization replete with shanty self-help housing juxtaposed to the private–public funded industrial parks where migrants worked resulted in the uneven geographies characteristic of Tijuana. The social goal of PRONAF was to inculcate a sense of national pride and identity into the border population. The reality of losing half of the nation's territory to the United States intervention in Mexico, as it is taught in history books in Mexico, remains an important lesson to Mexico and is remembered quite differently by Mexican career politicians than it is by their counterparts north of the border. This historical occurrence also provided impetus to the federal government in Mexico City to implement PRONAF.

In Tijuana, the project involved the creation of a new border facility as well as the city's first urban master plan geared toward the Tijuana River area. It included the removal of 1,500 families from an irregular settlement in the Tijuana River floodplain, called Cartolandia, because the homes were made from recycled cardboard and sheet metal, the construction of a major flood-control canal, and new boulevards and streets.

The renovation of the River District in Tijuana sought to encourage investment in this sector and provide a welcoming modern corridor for people entering the city from the United States. Architect Pedro Moctezuma accomplished this plan in the mid-1970s, and it is now the political, cultural, and financial center of the city known as Zona Rio. It is considered a distinguished district of Tijuana, where the main mall, upscale restaurants, fashion boutiques, and banking and government centers are located. Equally important, is the construction and eventual inauguration in 1982 of the Tijuana Cultural Center in the Zona Rio. This is an important arts center that combines a permanent museum, traditional gallery spaces with performing arts stages, botanical gardens, a small aquarium, and an IMAX theater housed in an iconic dome-shaped structure called *La Bola* by *Tijuanenses.*

This museum began as a social program to promote Mexican artists and cultural practices from the interior of the country to the border. As emerging vernacular art gained popularity on both sides of the border, the Cultural Center began to feature exhibitions of Tijuana artists. The museum frequently has border and international artists exhibit their works or perform at La Bola. For example, the Cultural Arts Center hosted many artists during the binational public art festivals called INSITE, which has public art and performance art produced by Mexican, United States., and international artists with installations on both sides of the border (*http://e-flux.com/announcements/31293/insite*).

The founding and progress of La Bola as a site for border artists also helped spur the burgeoning art movement that has become well established in Tijuana. *La Casa de La Culura de Tijuana*, a cultural center that preceded La Bola by 7 years, in tandem with La Bola, have become centers of artistic activity (see Figure 4.5). *La Casa de La Cultura* provides classes, workshops, and studios for burgeoning artists, which has led to the establishment of distinctive private art galleries found in the cross streets between Revolution and Constitution Avenues in the center of the city, such as *Pasaje Rodriguez*, an alleyway in downtown Tijuana with galleries and boutiques (Simmons, 2023). Other local artists' studios and music venues have opened in the Zona Rio such as *La Caja Arte y Cultural* (*www.lacajagaleria.com*) that features the work of *Tijuanense* artists, and *Le Conteiner* (*www.leconteiner.com*), a farm-to-table restaurant and local music-venue space.

FIGURE 4.5 *La Bola*, as Tijuana residents refer to the cultural center found in the Zona Rio neighborhood. Photo by David M. Walker.

*Glorietas*, or roundabouts, reminiscent of Reforma Avenue in Mexico City (itself designed to resemble the Champs d'Elysées in Paris) are found at the intersections in the Zona Rio district. In the middle of each *glorieta* are 40-ft.-tall statues of important Mexican historical figures, heroes, or iconic symbolism similarly found in Mexico City (see Figure 4.6 for an example). Moctezuma, Cuauhtémoc, and Benito Juarez stand proudly over intersections providing didactic symbolism to *Tijuanenses*: You may be less than 2 miles from the United States of America, but you are Mexican—*proudly* Mexican—and these figures (supposedly) struggled and died to ensure that you may live as a Mexican. Diana the Huntress also adorns a replica of a *glorieta* statue in Mexico City, while Abe Lincoln has another—for the role he played in aiding and arming Benito Juarez to expel the French and depose Emperor Maximilian in 1865.

## Border City Morphology

The renovation of the Zona Rio is an attempt to alter the morphological growth of Tijuana. Indeed, since the formation of the U.S.–Mexico international boundary, U.S. capitalists, Mexican government officials, and residents of Tijuana have constructed the urban built environment flush against the border. Because of Tijuana's relative location to Mexico City and the densely populated areas in the southern part of the country juxtaposed to its proximity to southern California, residents have used

FIGURE 4.6 The Cuauhtémoc Roundabout in front of the Luis Barragan-designed Hotel Camino Real. This form of didactic art displaying images of Mexican nationalism is a fixture in the built environment of Tijuana. Photo by David M. Walker.

the city's comparative geographic advantage to conduct business and remain close to the international border crossings. The city has grown hugging the U.S.–Mexican border, looking north as its urban footprint has spread from west to east from the Pacific Ocean. The most densely populated districts (called delgaciones) of the city skirt along la linea, facing north toward Southern California. This urban growth pattern differs greatly from the urban morphology of colonial cities in the interior of the country, or even historical Spanish cities of the U.S. southwest.

As many Latin American towns and cities (including those in Mexico) began to grow and develop during colonial times, their organization was mandated by a set of laws called Laws of the Indies, which dictated that each town and city must be built around a plaza. These laws were established by Spain to regulate the social, political, and economic structure of its colonies outside of Europe, and "mandated everything from treatment of the Indians to the width of the streets" (Griffin and Ford, 1980).

A typical plaza features the three pillars of colonial society that in Mexico would become the pillars of the newly found republic: the church, the federal government buildings, and the municipal government buildings. Mature colonial neighborhoods are adjacent to the main plaza. Currently, these colonial neighborhoods are in the process of transformation through government public–private gentrification programs, such as the *programa de rescate,* in Mexico City (see Chapter 2). As Mexican colonial cities grew, urbanization took the form of spines extending from the main plaza (or central business district) to markets and commercial malls. The elite residential sectors of the city are located along the spines, and along the periphery are the industrial parks and irregular settlements where the majority working poor live (Ford, 1996).

Mexican border cities in general and Tijuana in particular have grown very quickly with little planning. Infrastructure investment has been primarily targeted to the creation of tourist facilities and industrial parks. The "shanty," an organic architectural style limited to the periphery of Mexican colonial cities, is found throughout Tijuana. Shanties are built cheek by jowl next to high-income homes or even gated communities, and their juxtaposition reveals class hierarchies. Workers "invade," or squat on properties and construct their own homes that grow into neighborhoods within walking distance to maquiladoras in order for residents to be close to work. High-end spaces of consumption are adjacent to informal economic activities. Large-scale government cultural spaces compete with ephemeral workshops, galleries, and recycled-art spaces. Life doesn't center on the plaza in Tijuana as it does so often in colonial cities across the country (such as the Zocalo in Mexico City or the lively plazas of Oaxaca and Guadalajara City) where people across the spectrum of Mexican society congregate to celebrate daily life or major political events like *El Grito de Hidalgo,* or independence day. In Tijuana social life takes place at taco carts, parking lots, and a public park near Revolution Avenue called Teniente Guerrero Park. More and more, social life in Tijuana takes place in the private sphere of residences, party centers, and malls. The Zona Rio was an attempt to change this pattern and create a didactic built environment for Tijuana residents. The Zona Rio with its office buildings, banks, malls, and places of high consumption provides a stark contrast to the landscapes of the irregular settlement neighborhoods that cling to the ravines and hillsides adjacent to the river.

## The Border Industrialization Program and Urbanization

The construction of urban and industrial infrastructure initiated by PRONAF in 1965 began to change the urban morphology of Tijuana. The second federal investment program, called the Border Industrialization

Program (BIP), created infrastructure and tax incentives to attract offshore manufacturing investment from transnational corporations to the maquiladora sector.

*Maquiladora* is a Mexican vernacular term for assembly plants originally located in Mexican border cities such as Tijuana, but are now set up anywhere in Mexico. This export-oriented industrialization is an economic system of production that consists in the manual assembly of components in industrial workshops with cheap labor whose finished products are destined for markets in the Global North. The U.S.–Mexico border is one of the major areas of the world where this practice occurs.

In the maquiladora "twin-plants" system, a parent company has its research, design, and administration facilities in the Global North and sends equipment, supplies, machinery, raw materials, and other assets to company plants in Mexico for assembly or processing without paying import duties. The finished product can then be exported to the United States or to a third country also duty free.

Foreign companies operating in Mexico cut costs by using cheap labor. According to the San Diego Association of Governments (SANDAG, 2023), "Mexico offers lower wage rates than many Asian countries so the average company can save millions a year in labor costs." Equally important, Tijuana's proximity to the largest consumer market in the world cuts transportation costs. The city's location allows for just-in-time manufacturing, which keeps it competitive in an era of expanding transportation and communication technologies; it's still cheaper to drive or ship by train the assembled products 140 miles from Tijuana to the international airport in Los Angeles or the port of Long Beach than across the Pacific Ocean from China. This is good news for U. S. consumers. The flat-screen television sets and computer screens that are a ubiquitous part of life in the Global North would not be affordable to the average consumer if they were not assembled in maquiladoras in Tijuana, the television and computer screen capital of the world.

Maquiladoras provide employment opportunities and a higher income in Tijuana. The city has roughly 560 maquiladora assembly plants that employ 165,000 people. According to current data, a maquila worker can earn 600 pesos a week (or about $30 a week at the current exchange rate of 17 pesos to $1). Although this is not a large salary or living wage by U.S. standards, 600 pesos a week represents a higher wage than a worker is paid in the interior and especially in the south of the country. However, many of the consumer goods found in Tijuana are shipped from California, or residents cross the border to San Diego to make their purchases. Tijuana's site and situation in relation to food production means that an average consumer basket of goods costs much more along the border than it does in Mexico City or in other places in the center of the county. Beef is shipped from Sonora and Chihuahua, eggs and chicken from the United States, but milk and fresh vegetables and fish are local. This situation creates what economic geographers call a dollarized economy—even though the *maquila* workers earn more than their counterparts in the interior of the country, their purchasing power is less.

The composition of the labor force also has created sociocultural strife in Tijuana. Initially, 65–90 percent of the *maquila* workers were women between the ages of 16 and 24. According to *maquila* managers, women have the skills needed for assembly work. The operators falsely argue that women are more dexterous and patient and therefore can assemble the components for longer periods of time than men. Feminist geographers have shown (Cravey, 1998, 2012; Wright, 2014) that in reality, *maquila* managers preferred to hire women over men because they were theoretically perceived as easier to train and more malleable, as the assembly line may be their first paid job, and they are supposedly less likely to have knowledge of labor laws and therefore are less likely to participate in union organization. This thinking may have been true in the early days of the maquiladoras. Currently, women have become powerful labor organizers fighting for worker's rights, demanding safe and healthy working environments, and mobilizing government and corporations to meet their environmental responsibilities to clean toxic sites and follow environmental laws and regulations (See *Maquilapolis* in "Suggested Resources"). Nevertheless, it is along the U.S.–Mexican border in cities such as Tijuana where Mexican women first gained access to meaningful, well-paid jobs. The increase in access to paid work provides women with the power to make their own decisions, which can mean escaping abusive relationships, as they can pay for their own housing and make decisions on their reproductive care, deciding when and with whom to have children.

Industrialization and urbanization occur in tandem. We can see this in the United States where the country's early manufacturing centers of New York, Cleveland, or Pittsburgh witnessed rapid and uneven geographies of urbanization. In Tijuana, the growth of the maquiladora industry heightened continued uneven urbanization processes that have left a permanent trace upon the landscape.

In 1974, maquiladoras employed 9,276 workers, with women representing 79 percent. By 1986, the number had risen to 30,600, with women representing 77.7 percent (Profitt, 1994), a 300-percent jump in 12 years. Between 1986 and 2018, the growth has increased 600 percent. According to INEGI, or the Mexican government's official source of statistical information, in May 2020, Tijuana had roughly 760 maquiladora assembly plants that employ 190,000 people. By May 2020, the percentage of female workers in maquiladoras had dropped to close to 55 percent, which represents a higher participation in *maquila* work than women's participation in the workforce across the country. In 2023, women represented 46.8 percent of the total labor force in Mexico, including the *maquila* sector (World Bank Group, n.d.). This figure does not include the large number of women—as well as men—who work in the informal sector. (see Chapter 2).

## A Cosmopolitan Mexican City: Demographics and Economic Diversification

The rapid growth of *maquilas* from 1971 through the 21st century has had a ricochet impact on tertiary economies, population growth, the built environment, and gender relations/gender mobility. Interestingly, when considering population growth immediately following the initiation of the BIP in 1971, the rate of growth actually dropped from over 6 percent in the second half of the 1960s to level off to just over 4 percent throughout the 1970s. Throughout the 1980s, a stage of accelerated growth coincided with the Mexican debt crises; coupled with the fast-tracked growth of the *maquila* industry, the population growth rose back up to over 6 percent The population slowly declined from 6.34 percent in 1990 to 5.23 percent in 2000. In the early 2000s, the growth decreased by nearly half to eventually level off at just over 2 percent in 2018 and 1.83 percent in 2020. Numerically Tijuana's population has grown astronomically from 290,000 in 1971 at the beginning of the *maquila* era to an official population of 2.2 million by 2020 (see Figures 4.7 and 4.8).

The rate of natural increase in the population of Tijuana and other industrial border cities has historically been lower than the rest of the nation. This is because all around the world when women increase their education levels and enter the workforce, fertility rates decrease. Women are taking home their own paychecks, which changes their positionality

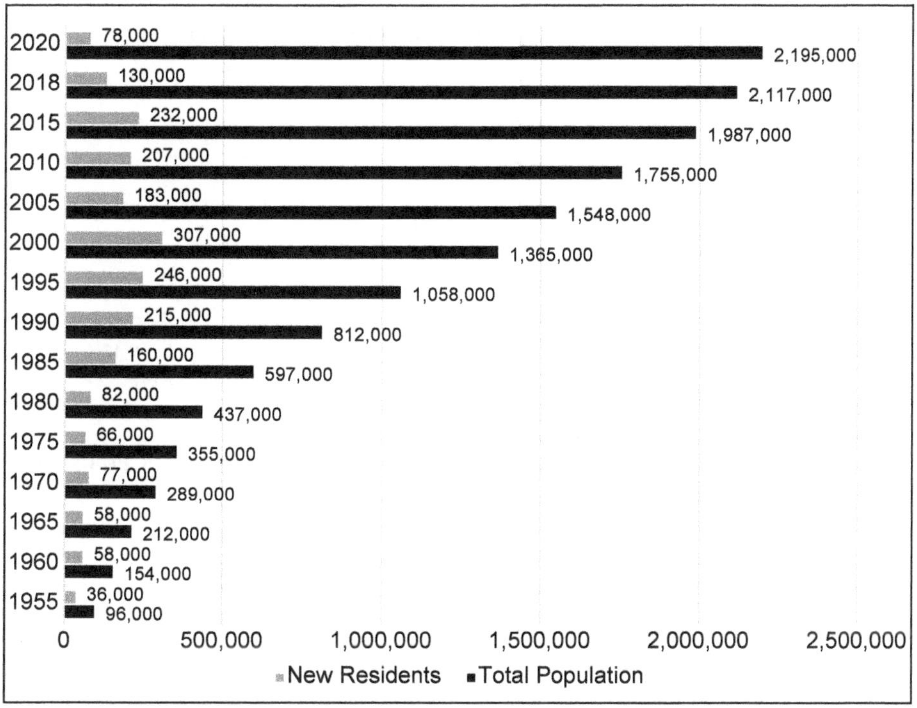


FIGURE 4.7 Tijuana's population increased in recent years due to a high volume of new residents. Chart by David M. Walker.

within the typical male–female Mexican family structure. They are making their own reproductive decisions, effectively empowering themselves through control of their bodies and thwarting patriarchy and *machismo*. Most of Tijuana's population growth is also driven by inward migration from the south. This migration has created a distinct urban culture in Tijuana that literally represents all states and regions of Mexico, with each group contributing the cultural idiosyncrasies that represent the *patria chicas* of Mexico. The growth of the maquiladora sector and the population of Tijuana has transformed it into one of the more cosmopolitan cities of the republic. Other border cities also house migrants from across the country, but Tijuana is the largest in population and has the most dynamic economy, including a dynamic arts and culture sector.

Similar to Tijuana's Mexican cosmopolitism, the tourist resort area of Cancun provides another example of a city that has attracted migrants

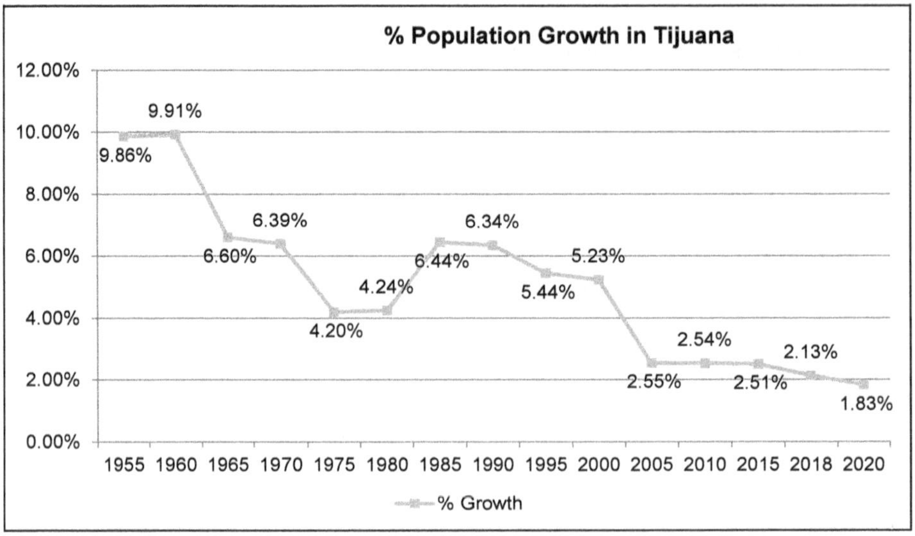


FIGURE 4.8 Tijuana's exponential population growth shown in percentages. Chart by David M. Walker.

from across the republic to labor in various industries. Cancun, a small Mayan fishing village in the 1960s, has been transformed into a tourism machine. The uneven geographies of development are similar to those in Tijuana: Cancun has a permanent population of nearly 700,000 people, 60 percent of whom live in irregular settlements consisting of migrants from across the republic, juxtaposed to resort landscapes that attract global tourists, and provides an example of urbanization similar to Tijuana.

## *Tijuanenses* Embody the Social Geographies of Music and Gastronomy

Tijuana's population growth coincided with the diversification of the city's economy from offshore assembly plants or the maquiladoras. From the 1990s, the growth in banking, higher education, and especially in the service sectors developed rapidly. The city's industry has continued to diversify. The well-educated population has been a magnet for start-ups (Hirsh, 2017). Food and culture—such as music and visual arts—have become noticeable engines of economic growth. The next section describes the social geographies of food and music production and how the representatives of these industries are producing new spaces of consumption at the border.

## Tijuana Music

Tijuana has often been romanticized by foreigners as being exotic, and its name has been used in the titles of songs. For example, Charlie Mingus's complete album *Tijuana Moods*, which the author claimed was his best work, aimed to recapture "an exciting stay in Mexico's wild and controversial border town" (Mingus, 1957). Herb Alpert, a Jewish American from Boyle Heights, Los Angeles, built an entire career with the Tijuana Brass Band, releasing 28 albums. And Manu Chao's song, *Welcome to Tijuana, Tequila, Sexo, Marijuana,* promotes the image of Tijuana as a vice city

To visit Tijuana is to be enveloped in music. Musical genres from across the country can be heard in the city. Buskers play on public transit, the sidewalk, or in front of taco carts. Musicians will go from restaurant to restaurant to offer their services. Tijuana has a small public space sandwiched between bars and food carts called Plaza Mariachi, where a cacophony of musical genres can be enjoyed. In addition to the musicians who provide the city with its quotidian soundtrack, more than 30 Tijuana musicians have achieved national stardom. For some, their fame has spread across Latin America, to the U.S. bilingual market, and occasionally beyond. For example, Tijuana No!, a ska, rock, and punk band influenced by The Clash, the Dead Kennedys, the Sex Pistols, and Black Flag, has gained notoriety across the Spanish-speaking world. Others, such as Julieta Venegas, who began with Tijuana No! and now composes and sings pop–rock–indie lyrics, or Exterminator, a popular *narco-corrido* band, have also captured audiences across Latin America and in the United States.

These three examples represent musical genres that emerged outside of Tijuana. A home-grown genre of music that is emblematic of the cultural syncretism at the border is *Nortec*, which is a blend of the words *norteño* and *tecno*. The genre combines electronic music with musical elements and instrumentation from *tambora* and *norteño* music. Its growth and popularity as a musical genre soon expanded to represent an artistic sociocultural phenomenon characterized by the amalgamation of local, national, and transnational influences while being distinctly representative of Tijuana in the 21st century. The founders of this genre are Jorge Verdin, who goes by the alias, Clorofila, and Ignacio Chavez, also known as Plankton Man, who started their careers as a computer programmer and art director, respectively. They wanted to capture the nocturnal sounds heard on Avenue Revolución in Tijuana. On one side of the street the sounds of electronic dance music pour out of a disco, while across the street people dance to *tambora* and *norteños* music. Clorofila and Plankton Man syncretized these genres, resulting in a music with its own sound. Voila! Nortec is born. They explain that it started as musical experiment based on loops and samples from northern Mexico (*tambora, corridos, banda sinaloense*, etc.) that were processed through analog and digital devices, and later became an interdisciplinary movement when graphic and fashion designers, video artists, writers, and even an architect joined the project. If you have been to Tijuana's Revolución Avenue, you have already experienced Nortec firsthand. Many high-profile television commercials have used Nortec as background music. The Nortec Collective's song, titled *Tijuana Makes Me Happy*, was used as the soundtrack for FIFA World Cup Germany and the video game EA Sports, demonstrating its popular global appeal.

## Tijuana Gastronomy

Tijuana's cosmopolitan gastronomy is produced by migrant–culinary entrepreneurs who arrived from the south but represent all regions of Mexico. Across the city entrepreneurs established small restaurants called *fondas*, which generally serve breakfast and a large meal for lunch and close early in the evening. *Fondas* attract workers who have the luxury to spend an hour sitting down for a meal with colleagues or friends.

More prolific across Tijuana's urban landscape are the taco carts offering regional tacos from across Mexico. Taco carts range from large, well-established, and permanent venues to small ambulatory carts that may either be clandestine or licensed by the city. A newly arrived migrant with good cooking and people skills can establish a taco stand with a moderate capital investment. If the *taquero*—a person who makes tacos—is especially adept at preparing the stews and grilled meats that go into the taco filling, she can attract a following and create a niche in the taco economy and may go on to establish a larger, more permanent taco stand. Many newly arrived residents will use the "taco economy" as a means of eventually expanding their enterprise or to get by until new opportunities materialize.

Many taco carts are located adjacent to the maquiladoras, government offices, and other places of employment, and near public transportation hubs. Tijuana is a busy city, and often people are eating on the run. Tacos provide a quick, delicious, and economic snack or an entire meal. As laborers leave work at the maquiladoras, they can grab a taco before running household errands, tending to sick relatives, providing child care, or entering to their second or even third jobs. The taco economy functions much as the informal economy of the street vendors in Mexico City (see Chapter 2). The production of tacos offers a livelihood for those who labor preparing and cooking the tacos and maintaining the taco carts, while providing an economical and rapid eating option to the working poor as they conduct their busy lives. The tacos may also provide a material and gustatory connection to consumers' hometowns in the interior of the country. For example, the maquiladora sector, where 35 percent of the economically active population works, consists of people who migrated from the south—from Jalisco, Tabasco, and Veracruz, for example—and from other parts of Mexico. To eat a taco like the ones they ate in their hometowns can provide a powerful sense of security and happiness. Finding food that is familiar can make Tijuana feel more like home and create a sense of place for the consumer.

Culinary influences brought by returning migrants from beyond the border to the city are noticeable in the foodscapes of Tijuana. Hamburgers, meatloaf, and the now world-famous Tijuana hot dog (wrapped in *beicon* [bacon] and served with mayonnaise, tomatoes, onions, and chiles) can all be traced to these returning migrants. Not to mention sushi, milkshakes, and imitation fast-food joints.

Many Chinese fled across the Mexican border with the U. S. implementation of the Chinese Exclusion Act in 1882 and influenced Tijuana's food spaces. While few original Chinese remain (Mexicali still has a large Chinese–Mexican population), the built environment in the form of restaurants with their distinctive pagoda decor still exists along the boulevards of the city. Many Tijuana migrant families continue to work in these Chinese restaurants.

Although as a culinary destination, Tijuana is perhaps best known for its street food, especially *mariscos*, *birria*, and tacos, the city also has a long tradition of fine dining. For example, one culinary staple that has diffused from Tijuana throughout the world is the classic Caesar salad. Here chef and writer Julia Child describes the salad:

> My parents, of course, ordered the salad. Caesar himself rolled the big cart up to the table, tossed the romaine in a great wooden bowl, and I wish I could say I remembered his every move, but I don't. They only thing I see again clearly is the eggs. I can see him break 2 eggs over that romaine and roll them in, the greens going all creamy as the eggs flowed over them. Two eggs in a salad? Two one-minute coddled eggs? And garlic-flavored croutons, and grated Parmesan cheese? It was a sensation of a salad from coast to coast, and there were even rumblings of its success in Europe. (Child, 1975)

Italian immigrant Cesar Cardini created the now famous salad at Hotel Caesar on Revolution Avenue in 1924 at the height of the "golden era" of vice tourism. His invention, and the popularity it gained, thus signaled that elegant, white-tablecloth dining has been at the center of Tijuana's food culture for many years.

Tijuana's food reputation has expanded beyond street foods, *fondas*, and even the tableside-prepared Caesar salad to include what numerous epicureans consider to represent the most innovative food prepared in the best restaurants found in Mexico, Latin America, and across the trans-frontier metropolis (Sanchez Macias, 2020). Tijuana chefs have taken local ingredients, elements from traditional Mexican dishes and home cooking and combined them with Mediterranean and Asian techniques to create distinctive culinary experiences that many are promoting as Baja–Med–Mexicano, or Alta Cocina Mexicano, a direct translation from haute cuisine or high-class cooking. Mexican chefs in Tijuana use the term to distinguish their restaurants from *fondas* or other establishments.

No better candidate represents this avant-garde food movement than Javier Plascencia who formed a partnership with his father in the Grupo Plascencia consortium that oversees 10 restaurants situated in upscale areas throughout Tijuana, including Casa Plascencia, a Spanish-themed meat emporium; Villa Saverios, the Roman-columned marble flagship; and Romesco, a Mediterranean-themed restaurant in San Diego. Located in the heart of Tijuana's Zona Gastronómica in the Zona Rio or "restaurant row" (see Figure 4.9), Villa Saverios is the go-to venue for birthdays and wedding showers among Tijuana's middle and upper classes.

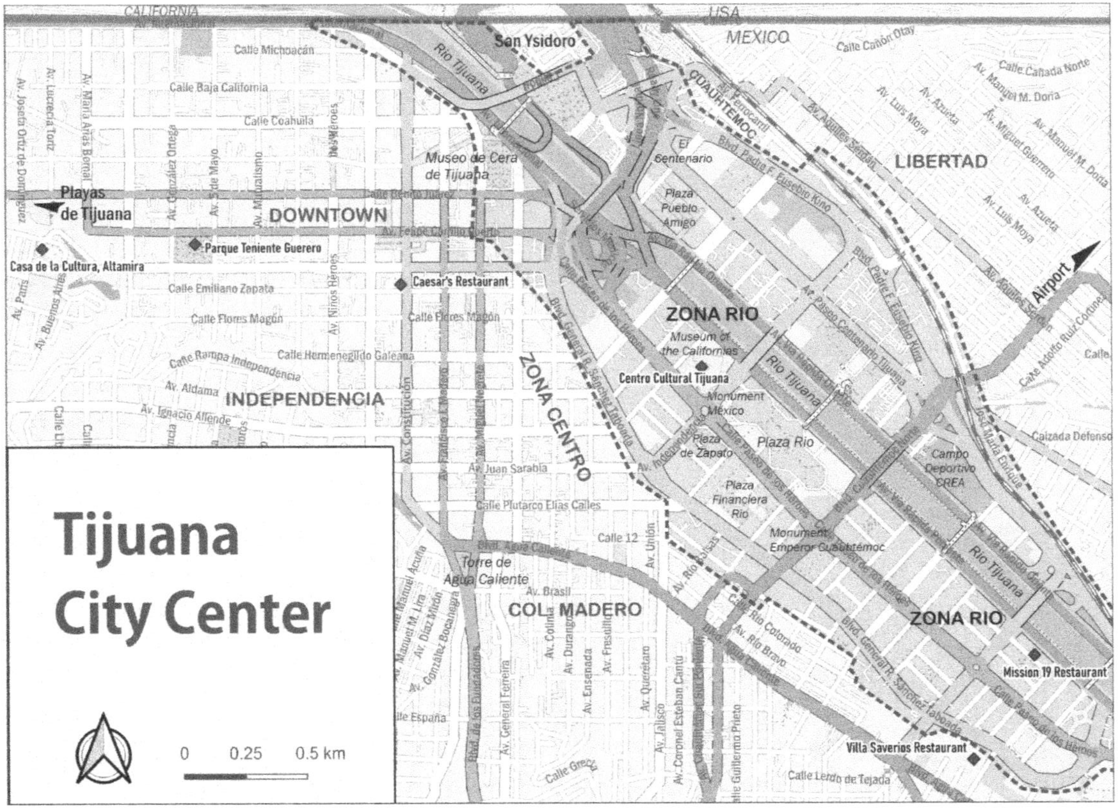


FIGURE 4.9 Map of Tijuana's Centro and Zona Rio, depicting the cultural center of Tijuana, the Casa de la Cultura, the Plaza Rio shopping mall, and Mission 19, Saverios and Caesar's Restaurants.

Plascencia's most ambitious restaurant, Misión 19, is located on the second floor of the country's second self-sustaining building constructed in Mexico. The postmodern concrete office building commands a 360-degree view of the city. There is no mistaking that you are eating in a Tijuana restaurant in the Zona Rio when you sit down to eat here. You can see the other postmodern buildings of the Zona Rio, the Bola, gaze across the international border and toward the east or west to view the ravines and barrancas that house irregular settlements, in situ housing, and the emerging postmodern apartment complexes that are now found peppered throughout the city (Brake, 2015). Mission 19's modern LEED architecture is replete with views of Tijuana's varied landscapes.

The story behind Javier and Grupo Plascencia's success didn't start at Villa Saverios or one of the other 10 upscale restaurants the consortium runs. It began like so many other stories in Tijuana: with a migrant who felt the push-and-pull of migration and chose to come to the city as it offered the best hope for success and a chance to work hard because "Tijuana is a noble city," as it has been described so many times, and here there exists the possibility to realize one's dreams.

Javier's father, Juan José Plascencia, now known as Don Tana, came to Tijuana from Guadalajara in the early 1960s. He first found employment in a maquiladora. In the mid-1960s the border between San Diego and Tijuana was much more fluid than it is today. There was no border fence or major surveillance. There was a customs house for cars, but pedestrians who had nothing to declare did not have to pass through the customs house, and eventually multiple organic paths were created by the foot traffic of people walking to and from the community of San Ysidro, which is now the far southern suburb of San Diego.

Tijuana residents could cross the border to purchase eggs and milk, go to school, visit family, or even get a pizza. Juan José did not stay long at the maquiladora and soon began working at a pizzeria in San Ysidro. He brought his newly learned skills back to Tijuana. He also brought his wife, Nana Chela, now famous for her culinary exploits, to whom Juan José is still married. She used her mother's pizza recipe in working with him to open the first Giuseppe's on Agua Caliente Boulevard in Tijuana. She originally stored the pizza boxes and cans of tomato sauce in the family garage. The pizza chain grew to more than 20 stores offering delivery of pizza á la Americana on mopeds and motor bikes that you can still see swerving in and out of traffic so the pizza arrives hot.

As so many children in Tijuana continue to do, Javier Plascencia went to high school and then culinary school in San Diego. He had grown up washing dishes and busing tables at the flagship Giuseppe's. There, he watched his father greet the customers as they came in through the door. He knew all their names, what their favorite dishes were, and when they celebrated their kids' birthdays. Javier learned the restaurant industry from his parents and codified his knowledge in the schools north of the border.

Javier opened his first restaurant in the Plaza Zapato, a hub of nightlife in the Zona Rio, in 1996. Frequented by evening revelers, it offered a small but well-chosen menu of Mexican classics and Spanish tapas. Javier spent his nights off frequenting new upscale European, Asian, and Californian restaurants across the border in San Diego and Los Angeles; he spent weekends in the food meccas of Mexico City, Oaxaca, and Puebla and went farther afield to Spain, France, and Italy. In these spaces of food culture, he honed his knowledge of Mexican, Asian, and Mediterranean food preparation.

Grupo Placensia's transfrontier consortium mirrors the urban forms of Tijuana that are produced in tandem by political, cultural, and economic processes from the United States and Mexico. The consortium combines the better aspects of gastronomy from both sides. Javier and his family are in good company in the production of elite consumption venues and the reinvention of *alta cocina Mexicana*. Cien Años, La Diferencia, La Querencia, La Espadana, and other restaurants are other high-profile spaces at which regional, national, and international cuisines are experienced, and they function as important sites of production and consumption, and reinvention of national identity at the border (Walker, 2013). These entrepreneurs provide a distinct layer of domestic-Mexican production of elite spaces that offers a contrast to the historical spaces of vice tourism where in the 1920s waiters at Hotel Cesar entertained Hollywood stars and Chicago gangsters while preparing their Caesar Cardini tableside salad. These Mexican haute cuisine establishments illuminate the importance of the taco economy in Tijuana and taco carts as sites of production and consumption for newly arrived migrants or third-generation *Tijuanense*. Javier Plascencia and his cohort contribute to these culinary layers of Tijuana honestly. They do not turn their back on the city that provided their mothers and fathers the opportunity to realize their dreams. Javier states, "I am proud of being from Tijuana" in a border English accent. And if pressed to describe what their favorite meals are, these women and men will take you back to where it all began in Tijuana.

## The Mexican Dream in Tijuana?

> "A mí no me llama la atención migrar a Estados Unidos, aunque sé que las oportunidades son buenas . . . es difícil prosperar en ese estilo de vida, tienen más estabilidad económica, pero no paran de trabajar, no tienen un balance. En mí no cabe el sueño americano. Prefiero irme a Puerto Vallarta, migraría a otro lugar en mi país." (quoted in Leon, 2018)
>
> ("It doesn't draw my attention to migrate to the United States, even though I know it offers good opportunities. . . . it's hard to prosper in that lifestyle, people have more economic stability, but they don't stop working, they don't have balance [in their lives]. The American Dream doesn't sit right with me. I prefer to go to Puerta Vallarta [on vacation, instead of the United States], I'd migrate to another place in my country [instead of crossing into California].")

This statement from 33-year-old Montserrat Bermúdez reflects a growing sentiment among well-educated Generation Z and millennials in Tijuana. She has what she considers a well-paying job as a receptionist that covers her rent and allows her to live in her own country. Her statement reflects the optimism that has emerged in tandem with *Tijuanenses* reclaiming spaces from rowdy drunk gringos and ubiquitous *narcotraficantes* (Leon, 2018). However, young *Tijuanenses* lived through rough times before and worked hard to reclaim their city and come to the realization that they wanted to make their lives on the southern side of the border.

Three factors negatively impacted Tijuana society from 2008–2011: the 2009 H1N1 virus, dubbed the Mexican swine flu, which halted the flow of day-trippers and revelers crossing from the United States; the economic crash in 2008 that led to an economic depression in Tijuana (when the United States sneezes Mexico catches pneumonia, as the adage goes); and an increase in drug violence. The upsurge of narco-violence in the city correlated with the United States exporting its war on drugs by pumping more than $2 billion into Mexico's drug war and discreetly deploying hundreds of operatives from the CIA, the Treasury and Justice departments and the FBI, as well as retired police officers and judges to Mexico during President Felipe Calderon's 6 years in office.

From 2009–2011, the convergence of these three factors turned Tijuana into a social ghost town. The economy gradually recovered and the spread of swine flu was controlled, but the narcoviolence kept people inside. In 2010 there were more than 800 homicides. The majority of those killed were rival drug gang members who aimed to fill the vacuum of the lucrative Tijuana smuggling corridor that was caused by the arrest and deaths of the Felix Arrellano brothers and the cartel's loss of power. The violence escalated to the point where bystanders were also killed in the crossfire.

Citizens marched in protests to plead with the government to intervene and thwart the violence. Calderon flooded Tijuana and Baja California with troops. Many protested the violence by not staying home, by purposefully going to art exhibitions, and by frequenting the few restaurants that remained open. As culture producer, writer, artist, blogger, and disc jockey Rafa Saavedra stated, "Tijuana is our home, our border roots, our web of friends, our work, and our dreams. Like so many who head out every morning to face the uncertainty and turmoil of capitalism gone wild, I can say this: I'm not afraid, I don't want to be afraid, I refuse to be afraid." (Saavedra, 2012). On the day that the bodies of six executed men were dumped at the playground of a school, Saavedra visited La Bola where he was happily surprised to see others taking his same cue: The place was packed with people.

The protests and continued participation in Tijuana social life perhaps eased the violence, but the subtle reality was that in 2011 the Sinaloa Cartel emerged victorious in this smuggling corridor, and the struggle over the profitable *plaza* ended. With a clear winner in the turf war, the violence has lessened. The drug smuggling continues because of the high demand in the United States, but with cartels not fighting with each other, the homicide rates along with the murders of innocent bystanders dropped. The foreign revelers did not return, which provided an opportunity for local *Tijuanenses* to construct social spaces solely for themselves, instead of reproducing pseudo-fantasy Mexico landscapes that catered to the U.S. appetite.

According to Genaro Valladolid, who was born in Tijuana and raised as a border crosser, the impetus of social change in Tijuana began at the legendary Dandy del Sur. Genaro went to school in San Diego, and after graduating from college worked for nearly 20 years in banks and brokerage offices in downtown San Diego, while living with his wife and raising their two children in Tijuana. During the recession in 2010, he was laid off from his job and got into the real estate industry, promoting and specializing in downtown Tijuana. "This whole movement, the multi-million-dollar projects we are seeing . . . was born in Dandy del Sur," Genaro insists (Saurez, 2017). "I said it . . . years ago [through a Facebook post], the people of Tijuana are going to make downtown alive again (ibid)," and he was right. Located on Sixth Street in downtown Tijuana, bar Dandy del Sur is one of the oldest cantinas. The bar had cachet with young *Tijuanenses*. "A lot of the creative class started hanging out there . . . they adopted it . . . the members of Nortec crew . . . writers like Rafa Saavedra . . . they started to create a scene (Suarez, 2017)", according to Genaro. The style was copied at other, older cantinas in the area, and new places began to open as well. Sergio Gonzales opened La Mezcalera, a bar featuring the agave spirit, connected to a disco called La MI-Ja, that caters primarily to the Tijuana creative class. Genaro believes that the opening of La Corriente Cevicheria Nais, a restaurant specializing in *ceviche*, solidified the movement toward the renovation-cum-gentrification of downtown Tijuana. Genaro explains, "The first people that started going to that restaurant were the hipster crew, the alternative. It eventually started to get more legit. The hipster crew brought their mainstream friends. Then bankers, office workers, and then moms and dads and grandparents." (Suarez, 2017). The renovation of Tijuana's once most famous restaurant downtown helped lead the shift in consumption spaces toward locals when the "Plascencia family took over Caesar's. The previous owners were being evicted. A close friend of the Plascencias went by and saw all the stuff outside, called them and said let's try to do something in there," explains Genaro (Suarez, 2017). The fusion of food, music, and art, with the opening of Pasaje Rodriguez, an alleyway reborn as a hip, artsy space with graffiti-style pieces, cool cafés, boutiques, and a brewery, created by people from Tijuana for *Tijuanenses* changed the spaces of downtown Tijuana where locals can perform their social geographies. These new spaces of consumption are perhaps becoming too successful, as they are attracting visitors from across Mexico and the United States. In 2022, the World Travel Awards nominated Tijuana as Mexico's "Leading City Break Destination" (Rivera, 2022). Does this accolade signify that the gringos are returning to Tijuana?

Locals are not only going downtown to eat and drink, hear music, and browse art galleries. People are moving downtown, and investors are taking heed. The real estate company, Cosmopolitan Group, owned by

David Saul Guakil, is best known for constructing the Torre Cosmopolitan, a postmodern nine-story building that earned a Platinum certification by the U.S. rating system Leadership in Energy and Environmental Design (LEED), has invested in downtown Tijuana. Guakil was born in Mexico City and migrated to Tijuana to build his fortune and raise his family. His two sons, born and raised in Tijuana, are architects trained in Mexico and the United States and contribute to the Cosmopolitan Group's success. The company's goal is to invest in projects that "rescue abandoned spaces in the city with the desire to modernize them and give them a new use" (Cosmopolitan Group, 2023). Building projects called Easy Living and Distrito Revolución were the first residential building constructed in downtown Tijuana in more than 50 years. The Esquina Sara building, also constructed by Cosmopolitan, is a downtown work-share space.

Investments in upscale living spaces downtown and across the city have continued in the post-Covid lockdown era. Cosmopolitan Group's competition, Guillot Arquitectos, completed five vertical upscale residential housing projects between 2020 and 2023 (Guillot Arquitectos, 2024). Accolades and awards for these new buildings are pouring in from Mexico City, the United States, and internationally. Indeed, the World Design Organization (WDO) nominated and approved San Diego–Tijuana as World Design Capital for 2024. The WDO awarded the cities for "their effective use of design to enhance economic, social, cultural, and environmental quality of life" (WDO, n.d.). The WDO provided this binational cross-border award—the first in its history, "as a result of their commitment to human centered design and legacy of cross-border collaboration . . . both cities are harnessing and mobilizing design to transform their region of 7.1 million inhabitants" (WDO, n.d.). The mayor of San Diego and Mexico's Secretary for Economic Development signed the official proclamation in an award ceremony where the mayor claimed, "this partnership between the design communities in San Diego and Tijuana is [a] demonstration of the connectivity and the interdependence of our two cities, especially when it comes to our position in the global economy" (Sloss, 2023).

Downtown restaurants and nightlife are attracting local hipsters and now again tourists. The building boom for upscale vertical housing is proceeding quickly and gaining awards and accolades. These changes in the downtown built environment and the verticalization of the Zona Rio accentuate the geographies of uneven development that have characterized the city. Just as during the vice tourism of the 1920s and 1950s, the PRONAF, the BIP, and the rapid growth of the maquiladora sector, the current development of the creative class economy and the built environment supporting this sector of the city's economy are juxtaposed to the irregular settlements where the majority of *Tijuanenses* live. The city is still lacking proper sewer and water connections. People are still pirating electricity. Maquiladoras still violate environmental laws, with renegade toxic runoff into the streams that flow into the Tijuana River and across the border to Imperial Beach in tandem with sewage from the irregular settlements. Indeed, every time the ephemeral Tijuana River flows it carries runoff from the maquiladoras and raw sewage. Due to this renegade flow, Imperial Beach remained closed every day from January through September 2023 (Letenyei, 2023). The city's population growth has outpaced the ability of the local government to provide the necessary infrastructure. The affordable housing that has been built is often hours away by public transport from job opportunities in the maquiladoras and the service sector. Guakil of the Cosmopolitan Group believes the city is facing its biggest crisis in the lack of transportation infrastructure improvements, housing, and sewage treatment plants. He believes that Tijuana can overcome these problems through innovation. Because of Tijuana's "geographic location next to one of the largest economies of the world, its people, its education centers, and its diversified economy" (El Imparcial, 2023), the city can overcome these obstacles. This effort will require hard work and partnerships among the people of Tijuana, combined with the local, state, and federal governments of Mexico in cooperation with actors from the city and county of San Diego, the state of California, and Washington D.C. to make it happen.

## DISCUSSION QUESTIONS

1 According to the Institute for Regional Studies of the Californias, 69 percent of residents in Tijuana are homeowners, while the rate of home ownership in San Diego is 58 percent. Criticisms concerning housing in San Diego tend to focus on affordability and the administration of property by landlords, whereas in Tijuana, respondents tended to complain about a lack of public services such as sewage and water connections. Based on the information in

the section " Social Geographies of Housing in Tijuana," (1) Why are the rates of home ownership higher in Tijuana? (2) Why do grievances concerning the housing situation differ between the residents of San Diego and Tijuana, and (3) How do the differences in complaints from residents on both sides of the border relate to the differences in home ownership?

2 According to the information from this chapter what is the taco economy? What do tacos represent for the people of Tijuana?

3 Decisions made by the federal government in Washington D.C. and in Mexico City influence the economies, social structure, and urban morphology of Tijuana. According to information from the chapter, name and describe one U. S. federal policy and one from Mexico City that impacted the built environment, livelihood strategies, and social geographies of Tijuana and its residents.

4 What would be the push factors be that would encourage people to want to leave their homes from rural Sinaloa, or rural Oaxaca? What are the pull factors that would entice people to want to move to Tijuana?

5 Describe the two types of land invasion highlighted in the chapter, how irregular settlements are constructed and transformed by *Tijuanenses*, and how irregular settlements become regularized and gain tenure.

6 How do border cities and interior Colonial cities differ in their urban form from one another? In your answer, explain how the urban growth patterns of Mexican border cities and Mexican Colonial cities produce distinct urban morphologies, and what are the reasons for the different urbanization processes?

7 Provide two reasons why fertility rates are lower in Tijuana and other border cities compared to cities in the interior of the country.

8 Describe, using examples from the chapter, why Generation Z and Millennial *Tijuanenses* prefer to stay in Tijuana and Mexico rather than emigrate to southern California.

9 Provide an example of how the renovation under PRONAF and BIP produced didactic symbolism in the built environment. Explain what didactic symbolism is meant to reinforce.

## SUGGESTED RESOURCES

Baja Window to the South. (2022). Tijuana's Community Building Renaissance with Genaro Valladolid of Bustamente Realty Group [Video]. YouTube. Available at *http://youtube.com/watch?v=6uW3JjFDXSg*.

For examples of transborder and transnational studies, see:

Publications from the Institute for Regional Studies of the Californias, housed at San Diego State University and El Colegio de La Frontera Norte in Baja California Norte at *https://irsc.sdsu.edu/publications*, *https://www.colef.mx/acerca*.

View the documentary *Maquilapolis* to learn about the professionalization of maquiladora workers, workers' rights, housing in Tijuana, and transborder *maquila* environmental contamination at *http://archive.pov.org/maquilapolis/film-description*.

For an excellent insight into the Mexico–U.S. border region, see:

Cravey, A. (2016). U. S.–Mexico borderlands In E. L. Jackiewicz and F. J. Bosco, (Eds.), *Placing Latin America: Contemporary themes in geography* (3rd ed.). Lanham, MD: Rowman and Littlefield.

For further readings on globalization at the Mexican border region and the impacts of the maquiladora sector, see:

Cravey, A. J. (1998). Globalization and its impact on the northern border region of Mexico. In S. Immerfall (Ed.), *Territoriality in the globalizing society: One place or none?* New York: Springer.

To learn more specifically about the maquiladora sector in relation to gender, see:

Peña, D. G. (1997).*The terror of the machine: Technology, work, gender and ecology on the U.S.–Mexico border.* Austin, TX: Center for Mexican American Studies.

Cravey, A. J. (1994). NAFTA, gender and the decline of the state in Mexico. In The International Geographical Union Commission on Gender and Geography (Ed.), *Beyond borders: The gender implications of multistate economic and social policies. Conference Proceedings* (pp. 60–89). University of Heidelberg.

## REFERENCES

Brake, A. G. (2015, September 7). House in Mexico by T38 Studio contains a private courtyard garden. *Dezeen*. Retrieved from *http://dezeen.com/2015/09/07/mexico-house-t38-studio-private-courtyard-garden-tijuana-black-stucco.*

Child, J. (1975). *Julia Child's kitchen*. New York: Knopf Doubleday Publishing Group.

Cosmopolitan Group. (2023). Impulsando El Dessarollo Y La Innovación en cada Protecto. Available at *https://cosmopolitangroup.mx.*

Cravey, A. J. (1998). *Women and work in Mexico's maquiladoras*. Lanham, MD: Rowman & Littlefield Publishers.

Cravey, A. J. (2012). U.S.–Mexico borderlands. In E. L. Jackiewicz and F. J. Bosco (Eds.), *Placing Latin America: Contemporary themes in geography* (2nd ed., pp. 205–218). Lanham, MD: Rowman & Littlefield Publishers.

El Imparcial. (2023, August 3). *Tijuana enfrenta los mayores retos de su historia: David Saúl Guakil. El Imparcial*. Retrieved from *http://elimparcial.com/tijuana/tijuana/Tijuana-enfrenta-los-mayores-retos-de-su-historia-David-Saul-Guakil-20230803-0014.html.*

Ford, L. R. (1996). A new and improved model of Latin American city structure. *Geographical Review, 86*(3), 437.

Griffin, E., and Ford, L. (1980). A model of Latin American city structure. *Geographical Review, 70*(4), 397.

Guillot Arquitectos. (2024). Guillot Arquitectos. Retrieved June 20, 2024, from *http://guillotarquitectos.com.*

Hirsh, L. (2017). Tijuana's dev shops target SoCal startup: Coding comes cheaper, closer and in same time zone., *San Diego Business Journal, 38*(42), 16–22.

Julia Child Foundation. (2022). Retrieved from *https://juliachildfoundation.org.*

Leon, V. (2018, July 8). El renacimiento de Tijuana, en el amanecer de la era AMLO-Trump. *Newsweek en Español*. Available at *https://newsweekespanol.com/2018/07/08/renacimiento-tijuana-amlo-trump.*

Letenyei, D. (2023, September 8). Imperial Beach in crisis: Sewage contamination threatens this California coastal community. *Green Matters*. Retrieved from *http://greenmatters.com/news/imperial-beach-california-sewage#:~:text=Sewage%20pollution%20has%20forced%20the,this%20year%2C%E2%80%9D%20Aguirre%20added.*

Mingus, C. (1957). *Tijuana moods* [Album]. RCA.

Proffitt, T. D. (1994). *Tijuana: The history of a metropolis*. San Diego: San Diego State University Press.

Rivera, S. (2022, June 6). Tijuana up for World Travel Award. *Border Report*. Retrieved from *http://borderreport.com/hot-topics/border-culture/tijuana-up-for-world-travel-award*.

Saavedra, R. (2012). *Tijuana dreaming: Art and life at the global border*. Durham, NC: Duke University Press.

San Diego Association of Governments (SANDAG). (2023, August 15). California–Baja California border crossing and trade data. Retrieved from *https://opendata.sandag.org/stories/s/Border-Crossing-and-Trade/adfp-wjae*.

Sanchez Macias, J. L. (2020, March). 8 restaurants in Baja California are recognized as the best in Mexico. *San Diego Red*. Retrieved October 26, 2024, from *http://sandiegored.com/en/news/181777/8-restaurants-in-Baja-California-are-recognized-as-the-best-in-Mexico*.

Simmons, J. W. (2023). Segregated economies in an integrated world: The gendered consequences of exchange rate movements in low- and middle-income countries. *International Organization, 77*(3), 668–669.

Sloss, J. (2023, May 25). San Diego–Tijuana honored as World Design Capital 2024. *Fox News 5*. Retrieved from *https://fox5sandiego.com/news/local-news/san-diego-tijuana-honored-as-world-design-capital-2024*.

Suarez, M. (2017). My seven years in Tijuana: A guide to understanding this changing city. *San Diego Red*. Available at *http://sandiegored.com/en/news/147700/My-Seven-Years-in-Tijuana-A-Guide-to-Understanding-This-Changing-City*.

Walker, D. M. (2000). *Ejido privatization and urbanization: Neoliberalism and landscape changes in Mexican border cities*. San Diego: San Diego State University Press.

Walker, M. A. (2013). Border food and food on the border: Meaning and practice in Mexican haute cuisine, *Social & Cultural Geography, 14*(6), 649–667.

WDO. (n.d.). San Diego Tijuana 2024. Available at *https://wdo.org/programmes/wdc/wdc-san-diego-tijuan-2024/#:~:text=As%20the%20first%20binational%20designation,legacy%20of%20cross%2Dborder%20collaboration*.

World Bank Group. (n.d.). In Mexico, the labor force participation rate among females is 46.8% and among males is 77.4% for 2023. Retrieved January 26, 2025, from *https://liveprod.worldbank.org/en/economies/mexico#:~:text=In%20Mexico%2C%20the%20labor%20force,labor%20force%20participation%20has%20increased*.

Wright, M. (2014). The gender, place and culture Jan Monk Distinguished Annual Lecture: Gentrification, assassination and forgetting in Mexico: A feminist Marxist tale. *Gender, Place and Culture, 21*(1), 1–16.

CHAPTER 5

# In the Dumps

## *The Social Geographies of Trash*

### Waiting on the Trash Man

At 6:00 A.M. in the working poor neighborhood called Iztaccihuatl in Mexico City, still-sleeping residents are awakened by the clanging of a bucket-sized bell announcing the arrival of a garbage truck. People hauling bags of trash come rushing out of their homes, dressed for work or still in bathrobes to meet the driver, present their "tip," and hand the bags of refuse over to men riding in the back of the truck. These men rip open every single bag and sort through the trash to salvage valuable recyclables before the refuse makes it to the dumps. The same scenario is taking place across town in the fashionable La Condesa neighborhood, where the trash is considered more valuable because the residents have a higher income, so it costs more for the garbage truck driver to obtain that route. When these garbage truck drivers finish their routes and end their workday, they sell the recycled materials to the bosses, called *caciques*, that run the transfer stations and dumps. According to Oscar, who is 60 and has worked in the garbage business since the age of 15, "as people arrive with their garbage it gets emptied in the bed of the truck or on the sidewalk where we separate everything [valuable recyclables like glass, copper, metals, aluminum cans, etc.]. We modified the garbage truck with hooks on the side where we hang large bags collecting the material we gather as it comes together in order to sell it . . . it is something we have to do to earn a little bit more money [to take] home" (Cruz Flores, 2023). At the transfer station the garbage is again sorted for valuable recyclables by informal garbage scavengers and the remaining garbage is taken to open-air dumps where it will once again be sorted. This is a very efficient way of recycling that is accomplished through the informal, the private, and the government sectors. As described by a female garbage scavenger in the municipality of Iztapalapa east of Mexico City, the center of garbage scavenging, "we have to sort the trash and be vigilant of what we scavenge and sell—that is our work, the work of the *pepenadores* . . . that's why I believe if they privatize [Mexico City and the federal government's plan to completely formalize and modernize the system of trash collection], we'll starve to death" (Cruz Flores, 2023).

The word used in Mexico, *pepenar*, colloquially means to pick through garbage, and is derived from the Náhuatl word *pepena*, meaning to choose or pick up off the ground. The men who work on the garbage trucks sorting trash, the street sweepers who also collect trash to sell to the garbage bosses, and the scavengers working in the dumps are all laboring in *la pepena*—picking through garbage. They effectively are all *pepenadores*.

Garbage pickers are found across Latin America. The vast terminology used to name this profession is not only witness to the size of the phenomenon, but also to the diversity of specialization within this trade: *cartoneros*, *buscabotes*, and *pepenadores* in Mexico; *churequeros* in Nicaragua; *basuriegos*, *cartoneros*, *traperos*, and *chatarreros* in Colombia; *chamberos* in Ecuador; *catadores* in Brazil; *buzos* in Costa Rica; *cirujas* in Argentina; and other terms like *recuperadores*, *recicladores*, *clasificadores*, *minaderos*, and *gancheros* (Fergutz et al., 2011; Marello and Helwege, 2016; Wilson, Velis, and Cheeseman, 2006). In English the terms *rag picker* and *scavenger* are used, and indeed, in cities from San Francisco to New York, scavengers, who are often recently arrived immigrants, modify small pickup trucks and roam the streets collecting cardboard. This carboard is then piled high in the bed of the modified truck and sold at recycling centers.

There are more than 25,000 people supporting their families by working informally in the private–public–informal consortium that is garbage collection in Mexico City. Fifteen thousand of these *pepenadores* live and work at the dumps of Mexico City (Martinez, 2018). Across the entire country more than 6 million people work in the private–public–informal economies related to waste (Urióstegui, 2019).

## Chapter Outline

This chapter looks at the shifting role that garbage plays in the urbanization of Mexico City. In doing so, it analyzes the correlation with the city's rapid postrevolution (1910) growth that accelerated with the implementation of import substitution industrialization (ISI) strategies and the increase in the amount of garbage (see Chapter 2 for an explanation of ISI). The increase in population not only produced more waste, but this period of modernization also coincided with more refuse consisting of inorganic and toxic materials. The cultures and economies of garbage collection during this period coincided with the garbaged networks of relationships between formal and informal economic activities and political client–patron relations among the *pepenadores*, the *caciques*/political organizers, and the ruling party from the 1930s until 2000, the PRI. As the country's political and economic structures changed in the 1980s and continued with the maturation of neoliberal policies well into the 21st century, garbage collection too became disciplined by neoliberal doctrine (see Chapter 2 for definitions of roll-back and roll-out neoliberalization). Informed by roll-back neoliberalization tactics, technocrats searched to maximize profit by changing the social and political structure of garbage collection in Mexico City. The aim of the current federal and state governments in Mexico City is to implement a public–private partnership between the state and private foreign recycling companies, with the goal of attracting foreign capital investments and ending the informal private recycling of waste by the *pepenadores*. The contemporary struggles in Mexico City over the production, collection, and processing of garbage involves conflicts between informal and formal actors that are played out in the urban arena of the city's streets, plazas, and garbage dumps.

## Who Are the *Pepenadores*?

Rag pickers have always existed since society began to produce waste. They exist in cities in every country of the world, including the United States. In Mexico City the number of *pepenadores* increased in tandem with the rapid postrevolution urbanization coupled with an increase in the production of garbage (Cervantes Nino and Hernandez, 2012). As open-air dumps began to emerge around the rapidly growing city, rural immigrants began to scavenge and settle in and around them collecting valuable recyclables that they in turn sold to earn a living. By the second decade of the 21st century, Mexico City had become the second-largest waste-producing megacity in the world after New York. The number of *pepenadores* increased in tandem. Entire extended families consisting of three generations live and work side by side sorting through the more than 13,000 tons of waste generated daily by the city's residents. There are people right now sorting garbage who were born at a dump, have labored in the dumps their entire lives, and will die there as well. Children as young as 5 work unsupervised. While many attend school, most only achieve a fifth-grade education as their labor is incorporated into the family structure (Cervantes Nino and Hernandez, 2012). Women constitute roughly 60 percent of the workforce at the dumps.

As methane escapes from the rotting trash, the waste is known to spontaneously ignite and create 30-foot-high walls of flames that can be seen from across the flat urban landscape of Mexico City. The *pepenadores* homes burn too; they rebuild them and continue to sort the trash. There is the story of a garbage boss, or *cacique*, "who controlled . . . (the) *pepenadores*, demanded sexual favors from the garbage pickers' daughters, and also took all (17,000) of his workers off to Acapulco on vacation once a year" (Guillermoprieto, 1990). Of course, there are also the reports of large rats weighing up to 15 pounds that are captured by the *pepenadores*. The *pepenadores* are exposed to the associated health and social risk factors that occur in spending their lives amid the trash. They cope with these circumstances. They choose this life of garbage picking and defend their livelihood, clashing with authorities to keep their jobs, demanding "*la basura es para todos*," garbage is for everybody! Why do they choose to live and work in the dumps? One family's trash is another family's livelihood, literally in the dumps of Mexico City.

Of course, it is people who produce trash. As populations grow and go through demographic transitions, consumption patterns also shift, which in turn produces distinct forms of trash. The following section describes how in less than a century Mexico shifted from a rural, agrarian society to a majority urban society. It is important to understand how this shift occurred, how it correlated with the production of more and different types of waste, and the consequences this change had on waste management in Mexico City. Population and waste production increases are linked to the story of garbage sorting in and around Mexico City as a metaphor for the shifting political economy of the city and country:

from clientelist PRI state and *caciquism* to privatization as part of the neoliberalization of society.

## Demographics and Garbage

Mexico's demographic changes in the first half of the 20th century included increases in population size in correlation with immunization and the growth of the health care infrastructure. Geographically it trended toward a rural-to-urban internal migration pattern (Snyder, 1966). In just as many countries across the Global South, immunization and advances in health care occurred before investments in education and job training, especially for women, which resulted in rapid population growth as the country shifted into stages 2 and 3 of the demographic transition. In Mexico, the sharp decline in mortality due to the expansion of health services, immunization, the low cost of importing effective drugs, and federal policies that sought to meet the labor needs of an expanding industrial sector and to populate the national territory, correlated with rapid demographic growth during the 20th century. Growth was especially rapid between 1954 and 1974, when the population increased by over 3 percent per year.

From the 1980s onward, the country introduced population policies that sought to reduce population growth. State-administered contraceptive programs, immigration to the United States that correlated with cultural influences on women's reproductive rights, and most important, increases in women's education levels and participation in the workforce correlated with a rapid decline in fertility during the last quarter of the 20th century. In the first quarter of the 21st century, women's reproductive rights have been at the forefront of changing family policies and shifts in political leadership, resulting in women obtaining reproductive rights in Mexico City, igniting the debate on a national scale (Amezcua, 2020). Mexico's population is expected, like that of the United States, Europe, Canada, Japan, South Korea, and China, to now reach stage 5 of the demographic transition. On the way toward that final stage, the rapid demographic growth of the past will continue to impact the age structure, as large cohorts born during the periods of peak growth pass through the life cycle. The transition will give rise to two changes, as pronounced as they are distinct: First, the population of working-age adults will grow rapidly; later, the number of older adults will increase rapidly, reaching, within the mid-21st century, aging ratios like those in developed countries, which may lead to a crisis of production and consumption on a global scale (Robbins and Smith, 2017).

The rural-to-urban migration trend directly following the Mexican Revolution (1910–1936) correlated with many peasants being pushed from the countryside due to the destruction of agricultural production and transportation infrastructure during the prolonged war. During the revolutionary period refugees sought subsistence in the urban centers of the central Meso-American plateau in Mexico City, Puebla, and Toluca. The rural–urban trend intensified during the ISI period with the rapid urbanization and industrialization of Mexico City, Guadalajara, Monterrey, and Toluca. With the advent of the emerging maquiladora sector and the federal border development policy, or PRONAF, described in Chapter 4, urbanization fueled by rural–urban migration from across the country led to population growth in the U.S.–Mexico border region (Ganster, 2017).

Today, 80 percent of Mexico's residents live in cities. More than 25 percent of the population resides in three cities: Mexico City, Guadalajara, and Monterrey. In less than 100 years, Mexico ceased to be a predominantly rural society based on subsistence agricultural production in the south and hacienda production and mining in the north, to becoming a major domestic industrial producer. Eventually the shift to offshore assembly industrialization in the maquiladora sector fueled the growth of border cities such as Tijuana and Ciudad Juarez. In tandem with industrialization and urbanization, Mexico shifted from producing primarily organic waste to toxic inorganic plastics, industrial, and chemical waste.

For example, just 30 years ago in Mexico City vendors sold street food wrapped in banana leaves, corn husks, or newspaper. They sold *horchata*, *agua de Jamaica*, and *tepache* beverages in gourds from which customers drank them on-site, not to go. Today food is all served in plastic. Even the beverages are sold in single-use plastic bags tightly tied with a straw sticking out the top for convenient on-the-go consumption. The country has begun to enter the stage of mass consumption dominated by the service sector. Contemporary capitalist economic systems function in stage 5 of the demographic transition to produce, or import, more goods for consumption, which in turn produces waste, garbage, refuse; the detritus of modernity in societies where perhaps we have never been modern (Latour, 1993).

The following section explains how trash becomes a commodity. It is important to conceptualize this transformation to understand why trash has become a multibillion-dollar-per-year industry, why *pepenadores* are fighting to protect their livelihoods in the streets and dumps of Mexico City, and why the federal and state government aim to formalize and privatize garbage collection and waste management.

## Conceptualizing the Value of Garbage

So how does trash become valued? What is the perceived value of garbage? Probably for many of us, garbage is something that is produced and consolidated in our homes. You organize the bins consisting of recyclables, maybe even compost, and then the regular trash that does not belong in these categories. Dead batteries and old electronics are disposed of responsibly, which often requires taking them to secure drop-off locations where you may be charged a fee. The garbage bins in your homes are lined with plastic bags, and once full are taken to the appropriate outside receptacles. For many U.S. residents, the garbage and recycling are most likely picked up once a week by a private contractor paid by the city to remove the refuse. The city then charges homeowners for trash removal. The consolidation of the waste industry demonstrates the value in trash. (An older generation may remember when municipalities took care of trash, and that being a unionized trash collector—or garbage man, as it was gendered work—was a job with retirement and health care benefits and vacation pay that allowed a worker to raise a family and send his children to college—but that is another story about garbage.)

But where does the regular garbage, recycling, or compost go when it is removed from your home? What do these private collection agencies do with it, and why are they in the waste-disposal business?

Globally garbage is a $580 billion a year industry (Allied Research, 2022). The private refuse collectors sell the recyclables to a domestic recycling plant where it processed and then sold in bulk to overseas or domestic manufacturing companies that reproduce the materials into new products such as recycled paper, boxes, glass bottles, cans, or even parts for autos, planes, and trucks, among others. The used batteries are shipped to, for example, Tijuana, Mexico, where the metallic lead and other valuables are removed, resold, and repurposed (see "Suggested Resources" in Chapter 4 for examples shown in the documentary *Maquilapolis*). The discarded and recycled electronics may be repurposed into the next smartphone, laptop, or video game console.

The organic, recyclable, or hazardous batteries and electronic refuse that people pay to have removed is correctly perceived as hazardous if kept in the household. When you look at your recycling bin, you perhaps do not see dollar signs. The refuse becomes a resource when it is taken from your house to the capitalist market sphere where it is sold and traded and effectively becomes a commodity. Hector Castillo Berthier (2003) conceptualized this process as, "one could begin by saying that garbage is worthless. When it is thrown into a wastepaper basket or bin it is worth nothing, but from the moment it is collected, transported, stored, classified, cleaned, sold, and reused, it is transformed into merchandise. This means its use value and initial exchange value can be recovered if human labor is incorporated. This can be expressed in the following formula: Garbage + Labor = Merchandise, which although apparently simplistic, implies a long, complex process of economic circulation" (p. 36). This resource/commodity is reconfigured as a usable item on the market that reenters the capitalist commodity chain–network to be resold on the global market until it eventually will again become a thrown-away object, or garbage.

In Mexico City, this process requires the laborer component, the *pepenador*: "garbage is the link that allows the Merchandise Cycle to close: Production–Distribution–Consumption (Recycled Garbage)–New Production" (Castillo Berthier, 2003, p. 36). In Mexico City, this multimillion-dollar industry requires the informal labor of the *pepenadores*.

In Mexico City the collection of garbage and its transformation into a commodity are a complicated process that involves many actors. It is important to learn about this process to understand that many workers and their families will lose their livelihoods if the government is successful at formalizing and modernizing the garbage industry. This next section explains this process, a day in the life of Mexican garbage.

## A Day in the Life of Mexico City Garbage

Distinctly from how many of us would take out the trash from our homes to outside trash bins once indoor trash cans became full, in Mexico City, consumers who produce the trash only bring the trash out to the street either when they hear the clanging bell of the garbage truck or the swooshing sound of the street sweeper. They dump the bags into the truck, at which point the contents will be sorted by a team of four men who collect materials to be recycled.

The city boroughs supply the garbage trucks. Drivers must pay a $5,000 bribe to gain access to the vehicle, for which he and his team are responsible for maintaining. They must also pay a bribe for the garbage collection route. The routes where wealthier people live are more desirable. Rich people buy and throw away more expensive things, and produce less organic waste and more recyclables. Drivers pay more for lucrative routes. The residents know when the truck will arrive, and although they do pay a small municipal fee for garbage collection, they understand and are prepared to provide a small tip, or informal payment, to ensure the service. Only the driver earns a salary, roughly $400 per month. The other four-to-five workers riding on the back of the truck sorting the garbage are "volunteers" (as seen in Figure 5.1). They earn what they can collectively scavenge and sell to the recycling centers combined with the tips. These earnings are shared among the driver and the team.

**FIGURE 5.1** Volunteers sorting trash in the early morning, Mexico City. Photo by David M. Walker.

At the end of the route, once the money from the tips and the sale of what the team scavenged have been collected, the leftover garbage is taken to the transfer station within the city or directly to the dumps where it will again be sorted on mechanical strips, leaving the unwanted materials behind. Finally, what is left is taken to the dumps. Until 2011 the rejected materials were taken to the Bordo Poniente and Santa Catarina landfills, both of which were closed in 2011, where they would have again been sorted by *pepenadores* who earned their living scavenging for and selling the last 10 percent of recycled material. Today it is taken to dumps 40–75 miles away to the states of Mexico and Morelos (see Figure 5.2).

Garbage collection and waste management in Mexico City is a hybrid interaction between formal institutions and informal agents. The municipal boroughs supply the trucks and the Mexico City government oversees the transfer station, separation stations, and the landfill sites, and the volunteers and *pepenadores* collect and separate the trash. This hybrid system of garbage collection and waste management in Mexico constitutes an interconnected structure, a dynamic framework constructed in everyday social interactions, and a series of transactions that connect different economies, spaces, and relations. Within this framework, the labor of the *pepenadores* has made the system function efficiently.

The next section describes how the informal work of the *pepenadores* within this framework relates to informal street vending and even land development and tenure. The negotiation between the formal and informal sectors of Mexican society has played a major role in the 20th- and 21st-century social urban structure across Mexico.

### VIGNETTE: The Temporal–Social Construction of Value

Just 8 miles west of Mexico City's Centro Histórico, the Santa Fe neighborhood is situated in the Sierra de la Cruzes Mountains. Santa Fe's landscape contrasts with the 17th- and 18th-century palaces, plazas, and pedestrianized streets found in the Centro. This hyperpostmodernist landscape bulges with many of Latin America's tallest skyscrapers housing transnational corporations (TNC) and upscale housing units. The neighborhood boasts two malls: Santa Fe Centro, Latin American's largest, and the underground Santa Fe Garden, which can

(continued)

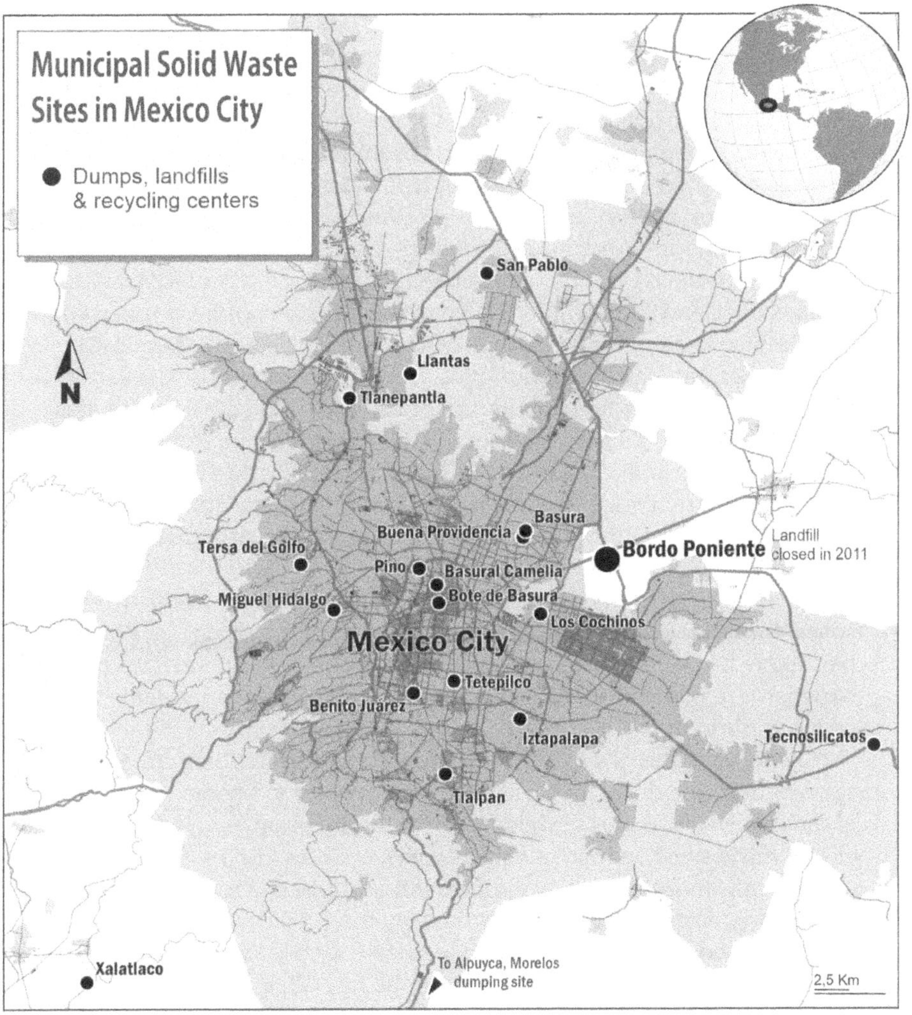


**FIGURE 5.2** This image shows a map of existing dumps, landfills, and recycling centers in and around Mexico City, the now-closed Bordo Poniente recycling center and landfill, the route to the Alpuyeca recycling center, and the dumping site in the state of Morelos. Map created by Dr. Georges Tadonki.

be seen from the street level through giant inverted glass domes, surrounded by green areas found within the medians among the automobile traffic. Contrasting with the traditional cantinas and fabled white-tablecloth restaurants in the Centro, here the foodscapes are dominated by *malinchista* eating establishments. In Mexico, a *malinchista* is a person who prefers and/or is infatuated with foreign, usually American, but also French culture, products, and people. The eating establishments in Santa Fe are *malinchista* because they are all foreign. The etymology of the word *malinchista* stems from La Malinche (Crassweller and Riding, 1985). The restaurants in Santa Fe consist of sushi restaurants, a P.F. Chang's—a gringo-imagined authentic Chinese restaurant chain owned by a New York-based private-equity firm, and Italian–Mediterranean trattorias. The automobile dominates the environment. The skyscrapers, TNC headquarters, mall spaces, filled with American, Chinese, European, and Japanese businessmen accompanied by their Mexican elite counterparts, are surrounded by rushing eight-lane traffic corridors throughout the day that become eerily quiet as night falls.

This is Mexico City's global transnational command center: an elite space embedded with multiple policies, practices, and processes of globalization. It has the most expensive office spaces and some of the most expensive residential spaces in the city—and all of this grandeur is built on top of garbage.

Land use changes in Santa Fe have shifted from being an area of sand-mining activity at the turn of the 20th century to becoming an area of shantytowns and garbage dumps from the 1950s onward, which coincided with Mexico coming into the modern era in tandem with the crises of modernity that accompany the process: garbage shifting from organic to inorganic matter, especially plastics. By the 1980s as the upper-class neighborhoods across the basin in Mexico City became saturated, the elites moved into the hills surrounding the city. This strategy aimed to separate the elites from the working poor and to escape the air inversions and contamination found in the basin. The lands around Santa Fe were privatized and purchased by developers who, in tandem with the CDMX (the city's government structure), created a self-contained, elite enclave. What was perceived as undesirable property due to industrial sand-mining activity that effectively became a garbage dump would eventually be reconfigured as a desirable place. Through neoliberal public–private partnerships, the elite were able to produce a high-tech, postmodern urban space that the rest of the urban Mexico City will never become: a high-profile corporate and residential elite enclave built upon gaseous garbage dumps.

This brief story of Santa Fe represents how value—in this case real estate/lived space—is socially constructed and shifts according to cultural practices and technical capabilities. But this chapter is not only about Santa Fe (see Valenzuela, 2007). The change in Santa Fe's perceived value from industrial site to garbage dump to elite spaces of production and consumption provides an example of how garbage as a commodity is also socially constructed and is transformed according to who is evaluating it; where the garbage is being produced, processed, and transformed into new forms of production; and how these processes are intertwined with social and material spaces. How garbage is deemed to be either refuse or a resource in Mexico City is imbricated in social and cultural practices, capitalist production, and changing political structures.

## *Pepenador* Efficiency, the Informal Economy, and the Social Structure of Urban Mexico

Mexico City's population growth occurred faster than the city's ability to provide housing and urban infrastructure, including garbage collection (Solíz, 2017). As we learned in previous chapters, rapid 20th-century urbanization in Mexico led to innovation in the informal sector, such as the construction of self-help housing and quotidian negotiations with the city and state for land regularization in Tijuana. In Tepito the working poor created the economic informal sectors of production and consumption discussed in Chapter 2. Within the context of the crisis of modernity in Mexico City—that is, consumerism and the production of trash—the working poor socially constructed refuse into a resource. The informal *pepenadores* organized around a boss or *cacique* became an efficient answer to recycling solid waste and processing garbage as Mexico City grew to 20 million people by the end of the 20th century.

Sixty percent of the labor market in Mexico City works in some form of the informal sector. This means that the people laboring in *la pepena*, or selling clothing, or preparing food to be sold in the streets do not have retirement plans, benefits, or social security. Self-help housing built on invaded lands where individuals or groups occupy land, squat, and begin to slowly build their homes and negotiate with the municipality or state for regularization—explained in Chapter 2—also forms part of the informal market. These examples of the informal economy are woven into the urban social fabric of Mexico.

Informal work pays more than unskilled jobs at corporations in the service sectors of the formal market such as in a Mexican Walmart (Walmex) or the convenience store Oxxo—Mexico's equivalent to the U.S. 7-Eleven. As discussed earlier, the subject of garbage may be conceived as an environmental and health risk, an aesthetic problem, or as an opportunity to earn a source of income (Sembiring and Nitivattananon, 2010). For the *pepenadores*, informal work as recyclers provides them a livelihood in which to survive and even thrive in the city. Moreover, although they work in the informal sector of society, in that they do not have benefits like retirement plans or social security, they do have job security: If people continue to consume, they will also produce waste. Simultaneously the *pepenadores* efficiently recycle valuable materials and contribute to the reproduction of consumer goods with little value added to these products in relation to their labor. Amazingly, the *pepenadores* remove the plastics, cardboard, aluminum, other metals, batteries, electronics, and electric–domestic items—anything of value—leaving behind organic waste and toxic garbage from households, restaurants, taco carts, hospitals, clinics, and industrial sites. The scavenged, valued waste is then sold to the dump bosses or *caciques* who then resell it to recycling plants in Mexico City or Toluca in the states of Mexico and Monterrey to be converted into recycled usable material that is shipped to manufacturing centers around the world. At these manufacturing centers the waste will be converted into a product that will have a new life as a consumer good. Salvaging for valuable garbage, the work of the *pepenadores*, functions as a highly efficient mechanism for removing recyclables from trash and is a multimillion-dollar industry (Cervantes Nino and Hernandez, 2012). Indeed, the system of garbage collection and waste management in Mexico City functions more efficiently than recycling programs in the United States, for example. More than 90 percent of recycled materials are removed from the solid waste in Mexico City and across the country using *pepenador* labor. In the United States the recovery of recycled materials and their reintroduction into the market has dropped to 6 percent, nationally. Not only is the work *pepenadores* do very efficient, but it also saves the municipalities and state money and reduces the cost of raw materials for industry.

The political figure of the *cacique* plays an intermediary role between the informal work of the *pepenadores* and the formal work of the state. This next section describes the role of the *cacique* within the garbaged networks of Mexico City. It is important to learn about the relationship among the *pepenadores*, the *cacique*, and state actors to understand how this informal–formal/private–public structure emerged and functioned before learning why the state aims to partially dismantle it.

## *Pepenadores, Caciques*, Client–Patron Relations, and the Social Structure of Urban Mexico

"Mexico's system is patrimonial, and that means that it operates through concessions, from top to bottom. . . . The garbage-collection concession is granted to the Federal District's Sanitation Department. From that point on, a number of subsidiary choices have to be made: whose trucks are going to collect the wealthiest garbage—the residential-zone garbage, with its mattresses and wine bottles and discarded clothes—and who is going to drive those trucks, because, of course, these things are scavenged by the truck crew long before they reach the dumps. At each step in the process where there is money to be made, a concession is granted, and at the end of the line are the garbage *caciques*, who tie the whole system together securely, and declare that the garbage is in its place, that the city is clean, and that its politicians are even cleaner" (Gallo, 2009, p. 301).

The system of garbage collection and waste management that emerged in the 20th century occurred within the framework of the social structure of urban Mexico, which was developed during 71 years of PRI hegemony. The government public sector with its administrative bureaucracy and state-owned industries controlled the labor sector, consisting of the Confederation of Mexican Workers, the National Confederation of Campesinos, and the Federation of Trade Unions of Workers at the Service of the State; the domestic private sector—usually consisting of former members of government who earned state contracts to build roads and factories, for example; and finally, the informal sector that includes the underemployed, self-employed, and informally employed with no social benefits or guarantees. Within this framework, the flow of resources determined power and allowed individuals to move within the system. Resources can be capital, labor, information, and just as important, political loyalty.

The resource that the *pepenadores* have is their labor and loyalty. Politicians and the *pepenadores* need a liaison, or a broker, to negotiate how the resources will flow. This broker more than likely emerges from a group of *pepenadores* and through charisma, strength, hard work, and by maintaining hardworking, dependent, and close-knit networks built on back-scratching and friendship—called *compadrazgo*. The political boss/*cacique* finds someone already embedded in the political system to provide him with connections and the chance to move vertically through the system. The relationship among *pepenadores* and *caciques* consists of the garbage networks that bring together the informal work of the *pepenadores* and the formal state apparatus. The *cacique* is a political and cultural go-between amid superiors and subordinates who has successfully articulated different political cultures and created connections between them. Thus, the *cacique* is astute at code-switching and is socially comfortable with both the political elite and the working class. Aspects of client–patron relations involving *caciques* in the three-tiered political organization of the Mexican state can be seen at the federal, state, and municipal government levels.

Strong man and political boss, *cacique* Rafael Gutiérrez Moreno—alias the King of Garbage—maintained dominance over the *Organización de los Pepenadores de México* from the 1950s until his death in 1987 (Treviño, 1993). Moreno fit the definition of an urban *cacique*. He began his career as a scavenger and garbage truck driver. His *padrino*, or nonkin networked godfather, Benjamin Carpio, became head of the Office of Waste Management in 1962. This put Moreno in a position to become manager of an important dump called Santa Cruz Meyehualco, where he became more involved in the official organization.

The PRI power pyramid benefitted from and rewarded his services and tolerated his activities. Moreno became a money lender to the general public and small businesses while donating large sums to the election process. He also ensured that the *pepenadores* from across the dumps would support the PRI during the election process through votes, political rallies, and union busting. Moreno proved skillful in negotiating the spaces of the dumps and the spaces of authority in the halls of government in Mexico City through brutal force, shrewd negotiation, and a mobilization of *pepenador* social identity and community. He placed large billboards adjacent to *pepenador* communities visible from transportation arteries that proclaimed, *Pepenadores son Mexicanos!* He built soccer fields, cantinas, and restaurants in the dumps for his workers. He knew when and where to provide rewards and gifts. With the money he donated combined with the political loyalty to the PRI by the thousands of *pepenador* voters he assured at each election, Moreno wove strong ties with the national PRI. This won him major status with the leadership of the PRI for which he was awarded by being chosen by the national PRI to run as a federal deputy, a position he easily won at the election polls (Cacho, 2014). Provided Moreno continued to donate money to the PRI and ensure the *pepenador* vote for the party, the party allowed him to control his garbage empire with impunity.

Moreno's son, Cuauhtémoc Gutiérrez de la Torre, became the garbage *cacique* after his father's death. He implemented the same practices as his father—donating money to the national PRI and ensuring the *pepenador* vote at election time. He became a PRI deputy that kept him in power well into the first quarter of the 21st century. He began to lose absolute authority over the dumps with decentralization in the late 1990s, (Cacho, 2014), eventually losing power in 2021 when prosecutors froze his bank assets after de la Torre was accused of running a brothel in the national PRI headquarters building in Mexico City and using PRI party funds to launch the business (Maldonado, 2021). Norma Gutiérrez de la Torre, a PRI official in Iztapalapa, and now head of the *Pepenador* Union replaced her brother. This story of the Gutiérrez de la Torre family, the *pepenadores*, and the PRI provides a classic example of *caciquismo*, client–patron relationships, and the collusion of the state with the informal markets.

Over the last 25 years, multiple political parties have successfully challenged PRI hegemony at the municipal, state, and federal level. The *pepenadores* fit neatly into the old PRI system, under which the party co-opted nearly every social group through patronage, thereby guaranteeing its hold on power. How much the rupture of PRI hegemony also eroded the framework of the social structure of urban Mexico has yet to be determined and is too large a task to be considered here. However, the introduction of neoliberal strategies for privatizing industry across the country has also seeped into the waste management system.

## The Neoliberalization of Garbage

With the advent of the debt crises in the 1980s followed by the initiation of neoliberal policies and the privatization of state functions—a process that continues through the 21st century—the Mexican political system under the PRI began to experience the devolution of power. This decentralization process ushered in a new political era, with local and state governments successfully challenging the 70-year dominance of the national PRI throughout the 1990s. This challenge to the PRI culminated with the National Action Party (PAN) party winning the presidential election in 2000. In the same year, the Party of the Democratic Revolution (PRD) won the mayoral race in Mexico City. The National Regeneration Movement (MORENA) party of Andres Manuel Lopez Obrador (AMLO) won the presidential bid in 2018, the governorships in four states, and the mayor's office in multiple cities, including Tijuana.

The activities of the *pepenadores* predate the implementation of roll-back neoliberal policies in Mexico City and these subsequent political changes. However, just as neoliberal technocrats have attempted to discipline the informal street vendors, or *ambulantes,* who produce and sell their wares on the streets of the historic center, so too do roll-out neoliberal policies seek to first take state control over the lucrative trash business, only to sell the industry of waste off to the highest direct foreign investor bidders and profit from the generations of *pepenador* work. Not only is the goal of the technocrats that of regularizing recycling in order to attract direct foreign investments and maximize the tax base, but there also exists the trope of modernization: that if Mexico City is to truly become known as a global city, then it cannot have thousands of intergenerational families sorting through garbage at the dumps. AMLO began formalizing urban space as mayor with the historic center's recovery plan. Marcelo Ebrard, who became mayor after AMLO, continued the process, while also installing security cameras throughout the high-consumption zones of the city and modernizing the buses with the introduction of the megabus. The removal of the dumps and formalizing the city's waste management process could prove to be his legacy. Finally, the argument is that if the state takes control over the garbage industry, it will be managed in a way that will be more environmentally sustainable.

The conflict over the privatization of garbage collection, with the goal of effectively taking the recycling business out of the hands (and reducing the power) of the *pepenadores* and into the coffers of a private–public partnership has raged since 2003. The *pepenadores,* who organized and became more professional through la National Confederation of Recycling and Industrial Waste (CONIMER), have negotiated with, held protests, and marched on the House and Senate, staged minor garbage strikes, and clashed with police in the streets of Mexico City to maintain control over the lucrative business of garbage. How did this public–private privatization strategy to gain control over the recycling industry from the *pepenadores* emerge? What directly caused the garbage conflict in Mexico City?

## Environmental Activism and the Closing of Bordo-Poniente Landfill

The Green Party—El Partido Verde Ecologista de México (PVEM)—is a political party founded in 1986 by Jorge González Torres. It is now headed by his son Jorge Emilio González Martínez.

The PVEM has become affiliated with anti-LGBTQ groups. A PVEM candidate, Gamaliel Ramirez, verbally attacked an openly gay candidate for Guadalajara mayor and called for enacting criminal laws against homosexuality. The party has been accused of nepotism, and has campaigned to reinstate the death penalty in Mexico. Two of its leaders have been implicated in working with the Romanian Mafia and Leticia Rodríguez Lara (also known as Doña Lety), the leader of the Cancun Drug Cartel. The party is not supported by Greenpeace, the international NGO, which considers the PVEM not to be an environmental party. The European Green Party does not recognize the party as part of its network.

In 2000 the PVEM made an alliance with the conservative PAN, helping bring to office the PAN candidate Vicente Fox Quesada, ending the PRI's 71 years of presidential hegemony. Interestingly, in 2003 the party switched its alliance to the PRI, reelecting PRI candidate Enrique Peña Nieto president in 2012, and currently, it has agreed to a coalition with the MORENA and PT in the Senate of the Republic.

In 2003, the PVEM proposed a modification of the law that allowed the federal government and Mexico City to oversee the management of all waste and recyclable materials, effectively ending control by the *pepenadores*. The party rationalized these changes with the rhetoric that a public entity would process the trash more efficiently, that the organic matter could be used in the production of energy, which would have environmental benefits. The implication was that the burning of methane gases produced by decomposing garbage could be used as a source of energy in Mexico City.

In 2011 the government of Mexico City petitioned the federal government to close Mexico's largest open-air landfill, Bordo Poniente. Bordo Poniente had grown to 370 hectares (about 914 acres) with over

70 million tons of garbage piled 17 meters (about 55.8 feet) deep. It had become one of the world's biggest open-air landfills. Just a year after the closure of the landfill, authorities announced that the contract for the closure of Bordo Poniente and biogas utilization had been awarded to BMLMX Power Company Sapi SA de CV. The consortium, consisting of two Spanish and two Mexican companies, signed a 25-year contract for the utilization of the biogas. It built an electric-power generating plant that sells power to the city for public lighting at lower prices than those currently offered by the Federal Electricity Commission. The consortium paid $2 billion to finance the decommission of the landfill and operate the biogas plant. Additionally, the government has constructed a composting site at the landfill, where it pays garbage truck volunteers and the *pepenadores* 50 pesos per ton to bring the organic waste to the composting plant.

*Pepenadores* protested the closing of the landfill, claiming that the closure destroyed their livelihoods and communities. Instead of bringing the waste to Bordo Poniente, the city hauls it to one of the 96 new dumps located across the states of Mexico and Morelos (see Figure 5.2). The current plan is to take the trash out of and far away from the city—just as New York City sends its trash to West Virginia or to float on barges offshore in the Atlantic Ocean. In the Valley Mexico and across the Meso-American Plateau, further out is not always an option because there are towns and cities across the region who don't want the city's waste.

This realization provoked the president of CONIMER, Francisco Urióstegui, to proclaim that within the pretext of climatic change, the PVEM, in collusion with state and private interests, saw the opportunity to enrich itself and further empower the privileged elites through public–private–corporate–controlled waste management. The initiative aims to remove the *pepenadores* from the dumps and construct 96 megaplants to attract direct foreign investment, to profit from the recycling of the materials on the global market, and to regularize and tax the industry—while also selling the methane gas, under the pretext of fighting climate changes.

The main worry of the *pepenadores*, who have been recycling waste for generations, is that any modification in the garbage industry could eliminate their participation in the commodity chain that transforms trash into a finished product. The president of CONIMER, Francisco Urióstegui, argues that the PVEM's initiative will drive the price of recycled materials higher. He is afraid that the government will seize control over the entire industry, sell the entire waste-management system to private firms who will then drive prices higher, which has happened with other industries in Mexico. In the end, he argues, the reforms the party seeks will not only hurt the *pepenadores*, but all consumers.

## The Professionalization of *Caciquismo*: La Confederación Nacional de Industriales de Metales y Recicladores (CONIMER)

The PVEM's proposal to change the garbage laws to give the state control over the industry represents the insidious manner in which roll-back neoliberal doctrine permeates civil society institutions perceived to be progressive. The rhetoric that the state can more efficiently manage the garbage industry is an argument contrary to the antipublic-sphere discourse employed in the Global North ever since U.S. President Ronald Reagan stated, "the nine most terrifying words in the English language are 'I'm from the government, and I am here to help.'" It is mobilized by neoliberal capitalist interests employing the PVEM as a mouthpiece to conceal the intentions of the state to sell the garbage industry to the private sector. So too has the disciplinary neoliberal doctrine professionalized CONIMER. Unlike Moreno, the *cacique* boss who controlled 15,000 *pepenadores* in the dumps of Mexico City, CONIMER has 96,000 affiliate organizations and representatives in 26 of the 32 Mexican states, representing more than 5 million informal garbage workers. This is no longer a form of 19th- and 20th-century *caciquismo*. With the dissolution of the centralized top-down, one-party structure, the local state and municipal authorities gained more autonomy, which during one-party rule were often subordinated and controlled by the *caciques*. The multiparty redistribution opened a fissure within the power structure that has led to organizations such as CONIMER to function as a "*federalización caciquista*" (*cacique*-like federalization), as was predicted years before (Castillo-Peraza, 1995).

The professionalized CONIMER employs the spatial tactics of protests, marches, and sit-ins at the Mexican House and Senate, while also forging alliances with neoliberally perforated political parties, such as represented by the cooperation between CONIMER and the Redes Sociales Progresistas, or the Progressive Socialists Network Party (RSP). With RSP's motto *#Sanar a Mexico*, or Heal Mexico, a broad-tent platform that includes gender and income equality and the promotion of sustainable growth while combating climate change, CONIMER aims to sway public opinion away from the PVEM. Its alliance with the RSP seeks to further legitimize their goals and to garner support for the *pepenadores* to maintain control over the dumps (Navarro, 2022). The CONIMER–RSP partnership has appropriated the language of free markets. In their promotional literature the parties employ the neoliberal doctrine of professionalization and expertise. They extoll the concept of the free market, with such statements as "the attempts by the government to take over the garbage industry is an incursion in an industry by inexperienced novices where there already exist expert business interests"; or "the Government is putting up obstacles to free competition in the market; the Government's objective is to monopolize the recycling market" (Navarro, 2022).

Will this strategy function to gain support for the *pepenadores* in Mexico City or, for that matter, from people across the entire country? As CONIMER's promotional literature states, "Mexico City is like a beacon that guides the country in the adoption of its laws" (Urióstegui, 2019), arguing that the city dictates what the nation will do. If CONIMER can't prevent the removal of the *pepenadores* in Mexico City, how can it protect the 6 million informal workers across the country?

In its current state, garbage collection functions as it has for decades. It is still a hybrid system with the municipal and state formal sectors working together with the informal-sector "volunteers" on the garbage trucks and, albeit fewer, *pepenadores* sorting the trash. Instead of piling up the trash in the city, the authorities choose to ship it out far from its points of inception. So now a day in the life of Mexico City garbage ends in a place like Alpuyeca, Morelos.

## Alpuyeca Does Not Want Mexico City Trash

Although the initial law introduced by the PVEM that aimed to change control of waste management remains in the courts, the dumps have been slowly closing across Mexico City. There is just too much trash to keep piling it up within the city and nearby urban areas. The city and its conurbation are saturated. Just as Santa Fe's land-use shifted from sand mines, to dumps, to corporate headquarters, so too will the preexisting dumps of Mexico City become developed into other uses. Furthermore, many of the 95 proposed recycling processing plants have been constructed, albeit, outside of Mexico City. One example is the village of Alpuyeca in the municipality of Xoxhitepec, in Morelos.

There are very few *pepenadores* at the dump in Alpuyeca. The trash is brought in at night and loaded onto large platforms. From the platforms the trash moves down conveyor belts where women stand on either side of the conveyor belt removing the valuable recyclable waste (see Figure 5.3).

The women work behind tall barb-wired fences. The trucks enter through large gates where security guards with holstered handguns and rifles in their hands stand by to let the trucks move in and out. No longer do entire families live and work together at the dumps. The women are

FIGURE 5.3 Women sort the trash at the Alpuyeca Dump. The women labor behind large barb-wired fences. This photo was taken from a 700-ft. distance. Photo by David M. Walker.

not freelance workers, informal *pepenadoras*, who have chosen to work in the dumps so they can work with their family members and earn higher wages than they can at Walmex and Oxxo (personal communication). These are salaried laborers. This form of garbage-picking production has been disciplined by neoliberal doctrine. Capitalism, with a capital C, has taken over the means of production. Why are there only women sorting through the trash on the conveyor belts? We can infer from Chapter 2 that the managers of the processing plant chose women under the pretext that they are more dexterous and therefore more adept at sorting through the trash for 8–10 hours per day. As we learned about the women working in the maquiladoras, managers hired them thinking they could pay them less than men, as for many it was their first paid job, so in theory they would be less likely to complain about working conditions or to organize unions. However, many women became labor organizers fighting for back pay when an assembly plant closed and demanded cleaner industrial sites, better working conditions, and child care.

There were a few *pepenadores* outside the gates where the platforms and conveyor belts were situated. They lived outside the gates sorting through trash that residents brought and dumped there—not the good garbage brought from Mexico City (personal communication). They were not able to get jobs at the processing plant, so they had to be satisfied with the meager offerings found among the refuse outside the gates, a far cry from the work the *pepenadores* had for generations in the dumps of Mexico City (Figures 5.4 and 5.5).

Alpuyeca is an Indigenous community. It has a beautiful, rustic early-17th-century church in the center of the village built on top of a precolonial pyramid. From the political messages and graffiti painted on walls around the community, it is apparent that the residents did not want the garbage dump, which opened in 2017 (see Figure 5.6). Now women, who were sorting trash with their families in Mexico City, do so at the Alpuyeca processing plant for an hourly wage. Young men live outside the gates. Unable to acquire employment at the plant, they must be satisfied with sorting inferior garbage discarded at the gates by local residents, alone, without their families. The Indigenous residents of Alpuyeca never wanted the dump in their community. Why is it there? Will it remain? Who is benefitting from this situation?

Alpuyeca is located within a 2-hour drive of large population centers of Mexico City, Puebla, Cuernavaca, and Toluca. It occupies an area of 34 sq. km. (about 13 sq. miles) and has a population of 8,330 inhabitants. Its

FIGURE 5.4 A *pepenador* living outside the gates of the Alpuyeca Dump. Photo by David M. Walker.

location is strategic, as the roads from the south and southwest converge there, and where the federal highways Mexico-Acapulco and Mexico-Grutas de Cacahuamilpa, the Highway of the Sun, and the intermunicipal highways cross it. Its geographical location makes this community an area coveted by investors. Their interests conflict with the interests of the community. For the inhabitants of Alpuyeca, their territory represents a guarantee of a subsistence economy. The village still functions as a rural community with small-scale agriculture and animal husbandry for local consumption. Because of its proximity to population centers and its rural way of life, Alpuyeca has become a desired site for the construction of megaprojects and investment, such as the transfer station where the trash is sorted and the open-air landfill and for investment in convenience stores like Oxxo and Walmex.

FIGURE 5.5 A *pepenador* dwelling outside the gates of Alpuyeca Dump. Photo by David M. Walker.

FIGURE 5.6 Political message painted on the wall in the Indigenous community of Alpuyeca: "Alpuyeca demands no to the garbage dump." Photo by David M. Walker.

As shown by the graffiti on the village walls (Figure 5.6), the residents are not passive about the changes in their community. Community members have denounced the plant, claiming that residual waste is flowing into the Apatlaco River, which they use to bathe and to water their animals. They have cited studies in which dumps have impacted the health of residents at other sites, listing skin diseases, leukemia and other cancers, and physical deformities suffered by children who grow up near dumps (Mendez, 2017). Alpuyeca is an impressive community as witnessed by its citizens preventing the construction of an Oxxo and continually protesting the dump. Alpuyeca's community members employ multiple resistance strategies, including requesting intervention from the National Commission of Civil Rights and the Environmental Protection Agency to investigate the abuses at the dump and help resolve the problem. "The government owes us an environmental debt" claims a community member, "it has built this waste management transfer center and an open-air landfill right in the middle of two rivers and a spring" (Mendez, 2017). Alpuyeca continues to resist the dump and other modernization efforts.

## Can *La Pepena* Be Included in Mexico City's Waste Management Transformation?

> The fact is that there are two or three things you can't mess with in Mexico City, . . . you can't touch the metro, the deep-drainage system, or garbage. Because, for better or worse, those things work, and the proof is that in this city, which is built on a lake, we've never suffered a major flood. But can you imagine what would happen if the sewer-system workers went on strike? It's the same thing with garbage . . . this is a clean city. But what would happen the day the garbage pickers shut down the dumps on us? Or if the truck drivers, most of whom have family ties to the pickers, went on strike? And you can't just solve the problem by removing the leaders; you have to find a way to replace them, or you'll have people killing each other just to get their own little garbage concession. We have to change things slowly; with the people we have. (quoted in Gallo, 2009, p. 307)

A technocrat and a Mexican politician educated at the University of Chicago School in neoliberal economic policy offered this opinion to a journalist. It encapsulates how complicated social urban structure is in Mexico City. The statement is also a glaring admission that there are spaces in Mexico City where informal activities function(ed) in tandem with formal state structures, and corruption is accepted in these spaces so long as they continue to work. Finally, also, the statement demonstrates that the politician admits that he wants the situation to be different, but due to the inertia of the social structure of urban Mexico City, the new society that the technocrats want to produce will emerge slowly, working *with the people we have*, which is a disdainful way for him to refer to his constituents.

Nevertheless, you can understand the politician's frustration about how slowly change comes about, as an anthropologist once put it in the title of her book, *Privatizing Latin American Garbage? It's Complicated* (Hill, 2015). The complicated hybrid structure of waste management signifies that simply referring to the changes in the garbage sector as a neoliberal privatization scheme—similar to how publicly owned industries, such as telecommunications, were privatized in the 1990s—is not accurate. Because, indeed, the garbage truck drivers and their volunteers epitomize the neoliberal argument that the private sector improves efficiency and lowers costs as they maintain their own budgets and pay themselves not an hourly wage, but according to what the market dictates the value of the collected goods are worth. This system incentivizes the garbage truck drivers and volunteers to collect more and to collect quicky, increasing their efficiency. In the current scheme of shutting down the dumps and moving them outside the city, there has been no demand to end the current form of garbage collection. Meanwhile, the *pepenadores* living and working in the dumps are losing their jobs and their communities. This is a problem, as many of the volunteers have family who are *pepenadores*, and as life gets harder for the *pepenadores*, the garbage collection unions may come to their defense.

Importantly, as the population of Mexico City and the surrounding towns and cities continue to grow, the authorities can't only be concerned with exploiting garbage for its marketable value, which is primarily the task of the garbage truck drivers and volunteers, and now increasingly, the *perpenadores* at transfer centers like Alpuyeca outside of Mexico City. The hybrid system of garbage collection functions well to exploit the value of recyclable materials. Mexico City and the federal government must create safe and sanitary waste removal that also protects and heals the environment.

The dumps that were shut down in Mexico City were not environmentally sound. Nor are newer ones built in the neighboring states, such as Alpuyeca's. New landfill construction must include lined beds to prevent ground water contamination and be plumbed to capture leachate and methane. Mexican authorities must implement preventive measures in the future construction of landfills to avoid contaminating the environment. In the long run doing so will prove more cost effective than, for example, the $2 billion cost of sanitizing the landfill at Bordo Poniente.

Forty percent of Mexico's population lives in the Central Plateau. To manage its waste, Mexico City needs to begin by first producing less of it. A ban on single-use plastics that took effect January 1, 2021, is a good place to begin. Also, citizens need to be better informed about the damage that single-use plastic has on the environment. In selected neighborhoods, the government of Mexico City has begun to provide garbage receptacles, called garbage islands, for residents to dispose of residential trash, instead of having them wait for the sound of the clanking bell and the arrival of the garbage truck. This change has led to complaints from the garbage truck drivers and volunteers who can't access the recyclables. Many of the receptacles are routinely filled with commercial and industrial waste from construction sites or informal manufacturing, not the intended residential waste. But this tactic may prevent people from piling bags of garbage in front of their homes, hoping that the volunteers will get to it before the dogs and rats.

These are all good things to do. Additionally, implementing a separation program at the point where waste is generated would be helpful in effectively managing Mexico City waste. Doing so could facilitate the sorting in the garbage trucks. It would also be more efficient to create specialized collection subsystems for each source—domestic, commercial, and industrial—of waste generation.

Nevertheless, aside from educating people to produce less trash (SEMARNAT, 2019), two processes need to occur to make waste management in Mexico City environmentally sound and equitable. First, modernize the landfills: Landfills should be built so that once the garbage has been picked over for recyclables and organic waste has been separated for compost, the remaining trash is compacted, and the landfill is lined to prevent groundwater contamination, covered with soil, and managed for emissions. However, the modernization of landfills in Mexico involves promoting the sale of equipment and technology produced in the Global North, especially in Europe in countries like Holland and Spain that specialize in incinerators, compost plants, metal foundries, sanitary fills, and so forth. European countries are aware of the opportunities. For example, the Netherlands Enterprise Agency commissioned a Waste Management Country Report: Mexico, with the "aim to provide a better understanding of business opportunities for the Netherlands in waste/circular economy sector in Mexico" (Huisman et al., 2021). There is lot of money to be made in the development of trash modernization in Mexico City as well as across Latin America. It represents another way of extracting money and resources from the Global South and siphoning them to the Global North. The $2-billion price tag for Bordo Poniente demonstrates this problem.

The second process is to include the *pepenadores*, garbage truck drivers, and the volunteers within this conversion to formalize and regulate waste management. They have efficiently mastered the *pepena*, have worked within this culture for years, and deserve to be included. It is the equitable thing to do.

Also, if the complicated sociopolitical dynamics of *la pepena* that are embedded in the social urban fabric of Mexico City and the hybrid nature of waste management continue, they may function as impediments to further democratize the city and the nation. Including the *pepenadores* and the volunteers in the formalization process of *la pepena* on the garbage trucks and at the dumps could eliminate the need for the *cacique*. The *cacique* is at the center of client–patron relations among the laborers and the state, and his adeptness at successfully negotiating informal and formal spaces hinges on the PRI-driven social structure of urban Mexico. The inclusion of *pepenador* labor could eliminate the *cacique*-centered corrupt system of kickbacks and bribes across layers of *pepenadores*, municipal employees, and politicians. It could also end access to a malleable and coerced voting bloc, easily mobilized to cheer at political rallies, bust unions, or disrupt student protests. Thus, creating a more transparent political process whereby perhaps politicians would be elected for their accomplishments instead of their connections.

The PRI certainly did not foresee this outcome when it controlled power in Mexico City and at the federal level for seven decades. With MORENA now controlling both branches of government, will this relatively new, leftist party change the social structure of Mexico City, or is this power structure—the institutions of the state, the unions, and the private and informal sectors—too embedded in the political and sociospatial psyche of Mexico? We can't answer this question now, but we can attempt to answer it by observing what happens at the dumps.

## DISCUSSION QUESTIONS

1 Go to your city or county website and research who removes the trash and recyclables from your home. Now, research the salaries and the medical and retirement benefits the companies responsible for trash collection offer to their employees. How does the process for trash collection compare to a typical day in the life of Mexico City garbage described in the chapter (i.e., is it privatized, is it run entirely by the city, or is it a hybrid arrangement, such as in Mexico City, and do home owners and residents play an active role)? How do the salaries, benefits, and working conditions of the refuse workers in Mexico City compare to those in your hometown or city? Now, using your school's library to access the local paper, do a historical search to learn when and if garbage collection and recycling were privatized in your town or city (if you can't find anything, choose a larger city in your state or region). How are the debates and issues related to the privatization of refuse collection you found in your research similar to or different from those highlighted in the chapter?

2 According to the information in the chapter, how does trash become valued? What is the perceived value of garbage?

3 How is garbage collection in Mexico City part of both the informal economy and the formal economy, or how does garbage collection, recycling, and storage in landfills employ both formal and informal workers? Please explain and provide examples from the text.

4 Explain how the cultures and economies of garbage, the garbaged networks of relationships between *pepenadores*, *caciques*/political organizers, and the ruling party from the 1930s until the 2000s represent a form of client-patron relations in post-Revolutionary Mexico.

5 How is trash related to demographics and modernity?

6 Aside from processing waste in an environmentally responsible way, what are additional preventative policies the Mexican state could implement to prevent the production of waste?

7 Based upon this and previous chapters, how is the privatization of garbage, with the goal to effectively remove the recycling business out of the hands (power) of the *pepenadores* and into the coffers of a private public partnership, indicative of neoliberal policies in Mexico?

## SUGGESTED RESOURCES

To learn more about the politics of garbage in Oaxaca, Mexico, see:

Moore, S. A. (2006). *The politics of garbage: Municipal solid waste in Oaxaca, Mexico* (University of Kentucky master's thesis). Retrieved from *https://uknowledge.uky.edu/gradschool_theses/288*.

No Waste Challenge website: *https://nowaste.whatdesigncando.com/mexico-city-en*

## REFERENCES


Allied Research. (2022). *Waste management market outlook.* Retrieved from *http://alliedmarketresearch.com/waste-management-market.*

Amezcua, M. (2020). El aborto legal en México tras una violación es un camino tortuoso. *El Universal.* Available at *https://interactivo.eluniversal.com.mx/2020/aborto-legal.*

Cacho, L. (2014, April 3). Cuauhtémoc Gutiérrez de la Torre. Mexican voices speak to the crisis of building their democracy. Retrieved August 2, 2021, from *http://elnorte.com/aplicacioneslibre/preacceso/articulo/default.aspx?__rval=1&urlredirect=https://www.elnorte.com/aplicaciones/editoriales/editorial.aspx?id=23648&referer=—7d616165662f3a3a6262623b727a7a7279703b767a783a--*

Castillo Bertier, H. (2003). Garbage, work and society. *Resources, Conservation and Recycling, 39*(3), 193–210.

Castillo-Peraza, C. (1995, 2 February). La yugoslavización del PRI. *Diario Reforma.*

Cervantes Nino, J. J., and Hernandez, L. P. (2012). Work in the informal pepena in Mexico: New realities, new inequalities. *Demographic and Urban Studies,* 27(*1*), 95–117.

Crassweller, R. D., and Riding, A. (1985). Distant neighbors: A portrait of the Mexicans. *Foreign Affairs, 63*(4), 920.

Cruz Flores, A. (2023, April 22). Aprovechan pepenadores un tercio de los desechos producidos en CDMX. *La Joranda.* Retrieved September 15, 2023, from *http://jornada.com.mx/noticia/2023/04/22/reportaje/aprovechan-pepenadores-un-tercio-de-los-desechos-producidos-en-cdmx-929.*

Fergutz, O., Dias, S., and Mitlin, D. (2011). Developing urban waste management in Brazil with waste picker organizations. *Environment and Urbanization, 23*(2), 597–608.

Gallo, R. (2009). *The Mexico City reader.* Madison: The University of Wisconsin Press.

Ganster, P. (2017). Binational cooperation and twinning: A view from the US–Mexican border, San Diego, California, and Tijuana, Baja, California. *Journal of Borderlands Studies, 32*(4), 497–511.

Guillermoprieto, A. (1990, September 17). Letter from Mexico City. *The New Yorker.* Available at *www.newyorker.com/magazine/1990/09/17/letter-from-mexico-city-3.*

Hill, S. (2015). Privatizing Latin American garbage? It's complicated . . . A view from the northern border of Latin America. *ReVista: Harvard Review of Latin America.* Available at *https://revista.drclas.harvard.edu/privatizing-latin-american-garbage.*

Huisman, H., Keesman, B., and Breukers, L. (2021). *Holland circular hotspot.* Waste Management Country Report: Mexico. Retrieved from *chrome-extension://efaidnbmnnnibpcajpcglclefindmkaj/https://hollandcircularhotspot.nl/wp-content/uploads/2021/02/Waste-Management-Latam-region_2021March-1.pdf.*

Latour, B. (1993). *We have never been modern.* Cambridge, MA: Harvard University Press.

Maldonado, C. S. (2021, March 20). La justicia acorrala al Príncipe de la Basura, otrora hombre fuerte del PRI en Ciudad de México. *El Pais.* Available at *https://elpais.com/mexico/2021-03-21/la-justicia-acorrala-al-principe-de-la-basura-otrora-hombre-fuerte-del-pri-en-ciudad-de-mexico.html.*

Marello, M., and Helwege. A. (2016). *Solid waste management and social inclusion of waste pickers: Opportunities and challenges.* GEGI Working Paper. Boston University: The Frederick S. Pardee Center for the Study of the Longer-Range Future Center for Finance, Law and Policy.

Marello, M., and Helwege, A. (2018). Solid waste management and social inclusion of wastepickers: Opportunities and challenges. *Latin American Perspectives, 45*(1), 108–129.

Martinez, G. I. (2018, March 12). Reciclan plástico y dan trabajo a pepenadores. *El Universal.* Retrieved November 17, 2018, from *http://eluniversal.com.mx/orgullomexicano/reciclan-plastico-y-dan-trabajo.*

Mendez, M. (2017, July 19). *Pobladores de Alpuyeca denuncian daños por planta de residuos sólidos.* Mexico City: CENCOS: Centro National de Comunicación Social.

Navarro, I. (2022, August 21). RSP busca transformarse en Movimiento Nacional Progresista tras perder registro. *Milenio. Politica.* Available at *http://milenio.com/politica/rsp-busca-transformarse-movimiento-nacional-progresista.*

Robbins, P., and Smith, S. H. (2017). Baby bust: Towards political demography. *Progress in Human Geography, 41*(2) 199–219.

SEMARNAT. (Secretaría de Medio Ambient y Recurson Naturales). (2019).

*Visión nacional hacia una gestión Ssustenable: Cero residuos*. Mexico City: SEMARNAT.

Sembiring, E., and Nitivattananon, V. (2010). Sustainable solid waste management toward an inclusive society: Integration of the informal sector. *Resources, Conservation & Recycling*, 54(11), 802–809.

Solíz, N. F. (2017). *Ecología política de la basura. Pensando en los residuos desde el sur. Ediciones*. Quito, Ecuador: Abya-Yala.

Snyder, D. E. (1966). Urbanization and population growth in Mexico. *Revista Geográfica*, 64, 73–84.

Trevino, A. C. (1993). *Basura de oro: El asesinato de Rafael Gutiérrez Moreno, lider de los pepenedores*. Mexico D.F. : Plaza y Valdés.

Urióstegui, F. (2019). La ciudad de México como un faro que guía al país en la aplicación de normas. Confederación Nacional de Industriales de Metales y Recicladores, CONIMER. Retrieved June 24, 2024, from *https://ejkrause.com.mx/camp17-green/bittecnica17/env7-FranciscoUriostegui.pdf*.

Valenzuela, A. (2007). Santa Fé (Mexico): *Megaprojects for a divided city. Revistas de la Universidad de Granada*, 40(1), 53–66.

Wilson, V., and Cheeseman, C. (2006). Role of informal sector recycling in waste management in developing countries. *Habitat International*, 30, 797–808.

CHAPTER 6

# Zoque Indigenous-Produced Space

## *Social Geographies of Los Chimalapas*

On a typically humid day in the remote jungle of Chimalapas, Oaxaca, Mexico's President Vincente Fox Quesada declared to the crowd of private Mestizo landowners, Zoque Indigenous farmers, Mestizo and Tzotzil Indigenous cattle ranchers from Chiapas, Mestizo forest workers, drug traffickers, and local and international NGOs that he wanted to be seen as "the President that had solved battles between brothers, the leader who had brought peace to the countryside in an inauguration of a new era of prosperity" (Gómez, 2004). This moment brought an end to the 47-year-old agrarian dispute between non-Indigenous cattle ranchers from Chiapas, who illegally invaded the Zoque territory and formed Cuauhtémoc Village, and the Indigenous Zoques who have long-standing claims and a connection to the disputed territory, believe they are the intended stewards of the Zoque jungle, and reside in the Zoque Indian villages of Santa Maria and San Miguel Chimalapas. Under the settlement, Cuauhtémoc land holdings were expropriated, the ranchers from Chiapas were paid a sum, and control of the village was placed in the hands of the Zoques, who reside farther away in what are identified as "headquarter villages," or *pueblos cabeceras* (personal communication). The highly publicized transaction fulfilled one of 14 campaign promises intended to place large tracts of land in the hands of an Indigenous minority with political power. Since then, the Chiapas ranchers have pledged allegiance to the

Zoques, effectively turning the Cuauhtémoc Village into a buffer zone aimed at protecting the Zoque lands. The Zoque elite encouraged the formation of buffer villages to prevent the invasion of their territory by illegal loggers, poachers, drug smugglers, and human traffickers from the neighboring states of Veracruz and Chiapas.

## Introduction

In this chapter we learn how the Indigenous Zoques in Chimalapas, Oaxaca, effectively created buffer villages around their ancestral lands to prevent the encroachment of illegal loggers, landless peasants, and also the NGOs from creating a biodiversity zone (Doane, 2017). This strategy allowed the Zoques to maintain autonomy over their territory, including the natural resources found there. The Zoques implemented material and metaphorical practices embedded in a larger framework that conceptualized space and nature to produce the defense spaces that the buffer villages represent.

## The Second Most Remote Region in All of Mexico

The municipalities of Santa Maria Chimalapa and San Miguel Chimalapa, which together make up the Chimalapas region, or *Los Chimalapas*, are situated within the Isthmus of Tehuantepec in the eastern corner of the state of Oaxaca that borders Chiapas to the east and Veracruz to the north. According to an NGO worker who oversees projects in the region, the Chimalapas is considered the second most remote region in all of Mexico as reflected through the absence of paved roads leading in and out of the dense jungle (personal communication). There is no water treatment plant, septic or sewer system, or hospital located within the two municipalities, only small, understocked health clinics operated by interns who come to the communities a few times a month from the state capital. The closest health facility is a 2- to 3-hour drive to Juchitán, the main commercial center on the Isthmus. The few schools that do exist are not registered with the national secretary of education and therefore unaccredited, making it difficult for the students who do finish high school to continue on to college. Many families therefore send their high-school-aged children to live and study with relatives in Juchitán. The municipality of Juchitán is located south of the Chimalapas and to the west is the region known as the Mixe. Located in a border zone between the pacific port of Salina Cruz and the gulf port of Coatzacoalcos, the Chimas, as the region is called vernacularly by neighboring settlement populations, is located at the meeting point of the Sierra Madre Occidental and Oriental Mountain ranges. Together San Miguel Chimalapa and Santa Maria Chimalapa reflect the largest communal land surface in the state, nearly 6,000 square miles, representing 10 percent of agrarian community lands in Oaxaca (Aguirre, 2015; de Teresa, 2000). These two large municipalities had a combined population of more than 15,000 inhabitants (INEGI, 2020), out of a total population of nearly 600,000 across the two Oaxacan districts of the Isthmus of Tehuantepec and Juchitán, and the Tehuantepec districts.

The Chimalapas is considered one of the most important biological areas in Mexico primarily because of the variety of ecosystems ranging from tropical cloud forests and savannas to evergreen high forests (Figures 6.1 and 6.2). The tropical forest is home to a greater variety of trees than in all of the United States and Canada combined (Stanton, 1992–1993). Within the remote interior live such threatened species as the *tepezcuintle*, a native jungle "pig" (it is in reality a large rodent that is considered a local delicacy), the jaguar, the spider monkey, the ocelot, and the tapir.

Environmental groups all over the world have rallied for the preservation of the Chimalapas owing to the region's potential to counter the effects of global warming (de Teresa, 2000; Umlas, 1998). The interest in the region by environmental groups, especially by international and local environmental NGOs and how they how they frame the Chimalapas as a "last rainforest frontier" that must be "preserved," has greatly influenced Zoque professionalization. This next section reveals the spatial and cultural distances that exist between the Zoques' lived experience in the jungle and those who work for a large NGO.

When you drive the 6 hours from the World Wildlife Fund's (WWF) offices, comfortably located in Oaxaca de Juarez's upscale Reforma neighborhood, to the Zoque village of Santa Miguel Chimalapas, located on the Isthmus of Tehuantepec roughly 95 km. (59 miles) inland from the Gulf of Tehuantepec and the Pacific Ocean, you pass through a spectacular array of differing landscapes, climates, and cultures. Leaving the Reforma neighborhood with views of the red-tiled roofs and bell towers of the colonial churches, you reach the eastern periphery of the city and begin to enter neighborhoods of disamenity. As you gradually climb out of the central valley of Oaxaca, you pass by the site of a 2,000-year-old Montezuma Cypress Tree at Santa Maria del Tule and the famous Zapotec rug-weaving village of Teotitlan del Valle before driving by the crossroads to the Mixtec archeological zones of Yagüi and Tlacolula. Once you pass by the rich agave fields where *palenques* (distilleries) produce mezcal in the villages of Matitlan and Mitla, you begin the descent to the Isthmus. The warm but dry heat becomes humid. The dry hills and fields of agave turn greener. This is no longer a land area where you hear Mixtec spoken alongside the Zapotec language of the central valley. The Isthmus cities of Tehuantepec and Juchitan are Zapotec-speaking communities, *Los bInnIZá* in Zapotec, or *el pueblo que proviene de las nubes*,

FIGURE 6.1 Forests in the municipality of San Miguel. Photo by David M. Walker.

FIGURE 6.2 An image of the forests in Santa Maria Muncipality. Notice the clear-cut logging, but also see where the Chimas community members have replanted. Photo by David M. Walker.

the people who descended from the clouds, in Spanish. Tehuantepec and Juchitan are Isthmus Zapotec-speaking communities that have continuously existed at these same sites for millennia, or time immemorial. The drive to the Chimalapas, where the Indigenous Zoque communities have lived for 6,000 years, takes you north across the salt flats of Laguna Superior, crossing the Isthmus until you reach the southernmost point of the Sierra Oaxaca and begin the ascent of Cerro Atravesado. As you climb the mountain with an elevation of 7,000 feet, you escape the humidity of the valley and smells of drying fish and shrimp along the Laguna and turn off the main highway, Carretera National 185, that crosses the Isthmus to Coatzacoalcos, Veracruz. The landscape shifts from wetlands to selva. As you drive along the dirt road, you pass through tropical, semideciduous cloud forests surrounded by rising peaks of pine forests and deep valleys. You drive along a dirt road for nearly 2 hours to arrive at the Zoque-speaking village. In 6 hours, you have traversed a multitude of landscapes, cultures, food and beverage practices, and distinct Indigenous communities. The small cinder-block edifice that houses the Zoque council, with open windows and no air conditioning or wireless internet, provides a stark contrast to the comfort of the WWF's office in the Reforma neighborhood of Oaxaca (see Figure 6.3).

On a warm and wet day in June, I took this drive with employees from the WWF who negotiated with the Zoque communities to create a biosphere reserve. As part of a binational research team studying the effects of globalization on NGOs in Oaxaca, I volunteered at the WWF's offices to learn about the process of acquiring funding and about how money was spent by international NGOs, WWF Switzerland NGOs, WWF Mexico, local NGO partners, and in the communities where these institutions worked, such as Los Chimalapas. We were headed to the Chimalapas to meet with the Zoque community, local and state government officials, and local NGOs, all of whom are stakeholders in the WWF's plan to create a biosphere reserve. This preliminary meeting, organized by the WWF, was not intended to create any immediate specific policies, but to learn about each stakeholder's interests in Los

FIGURE 6.3 San Miguel Municipal Auditorium, where the Zoque community members met with NGOs and government officials. Photo by David M. Walker.

Chimalapas. It was an amicable meeting (future ones would not be; see "Suggested Resource" at the end of the chapter). Perhaps 2,000 Zoques, eight representatives from the WWF Mexico, as many as 40 delegates and representatives from state and local governments and local NGOs, and me, were in attendance. The Zoques were proud to have us in their home—not in any one dwelling or even the village of Santa Maria, but in the Los Chimalapas region. As is the custom across Mexico, when you invite guests to your home, you prepare a meal. The meal—which was prepared and served outside and consisted of a stew of meat with red chiles, similar to a *pibil* sauce, nopales, tortillas, and water—was delicious. When the WWF team departed, the team leader, my *tocayo*, or namesake, David, stopped for a case of beer to relax after what was for him a stressful day to celebrate and talk about the day's event. We stopped halfway down the mountain to get out of the car. While David led the discussion, another WWF employee, Rafa, looked disturbed. Rafa came from Mexico City and had all the airs of a *capitalino*—meaning that he considered anyone who lived outside of Mexico City inferior. He also happened to be a vegetarian with a master's degree in biology. His biggest concern was what he had just consumed at the communal meal served by the Zoques. Rafa's colleagues at first ignored his question, which was quickly becoming a plea to learn what we had all eaten. Finally, a Oaxaca native and long-term employee of the WWF, Javier, blurted out, "Rafa, what you and all of us just ate was *tepesquintle*." Rafa was shocked and momentarily speechless. Although the *tepesquintle*, a spotted paca, or large jungle rodent native to southern Mexico, is not yet under the threat of extinction owing to overhunting—the meat is prized among Indigenous communities—because of development and the destruction of its natural habitat, its numbers are dwindling. When Rafa eventually recovered, he asked, "but aren't *tepesquintles* facing extinction," to which Javier responded, "Rafa, in order to protect endangered animals, you have to eat a few of them, and you must learn to eat endangered animals with the communities that are going to preserve them!", encapsulating the clash on the ground of the Zoques lived reality and Global North environmental idealism.

It is at meetings like this one organized by a prominent NGO, with smaller, local NGOs and government officials present, where Zoques become experts in the language and practices of conservation and preservation. The Zoques astutely combine what they have learned from the NGOs and government officials with their own knowledge of the boundaries and spaces of their territory to preserve their power and maintain control over Los Chimalapas.

A description is needed to understand the trajectory of Zoque population expansion into Zoque territory. The Zoques have deliberately created and populated villages to expand their territory (Figure 6.4). This steady growth has led to the creation of the buffer villages juxtaposed to the headquarter villages of Santa Maria and San Miguel Chimalapas.

## Reverse Settler Colonialism

First, a note on Zoque reverse settler colonialism. Zoque colonization did not involve an invasion in which the Zoques populated the lands of other Indigenous groups or created Zoque villages in the state of Chiapas. Reverse Zoque settler colonialism is an expansion into Zoque-held land—further populating the territory to which the Zoques had and still have claims. This form of colonialism is not the same as the land invasions by cattle ranchers from Chiapas and loggers from other states across the republic.

Historians understand Zoque colonization of the area as occurring in phases. Between 1800–1948, the Zoques left the municipal capitals and founded nine villages between the two *cabeceras*, or headquarter villages, as a result of a growing population (Gimenez, 1996). Initially, the men would venture out alone and establish plots for agricultural production. Upon returning to the headquarter villages, they mobilized families to colonize the central and southern areas of the Chimalapas. Five more "first-phase" villages were founded based on this model between 1935 and 1947. The second phase of Zoque colonization, which ended in 1995, was not a result of demographic pressure but was rather a response to infringements by government-sanctioned agricultural workers, licensed and clandestine logging companies, drug traffickers, and human smugglers. According to one Zoque resident, colonization by the Chiapas government, invasions from big multinational corporations, *mestizo* farmers, and drug dealers were the primary factors that have influenced the development of the "controlled towns" (personal communication). The Zoques invited outside members to settle the buffer zones to prevent unwanted colonization and established a governing elite that oversaw the administration of the new towns, simultaneously insuring fidelity and subordination to the villages of Santa Maria and San Miguel (personal

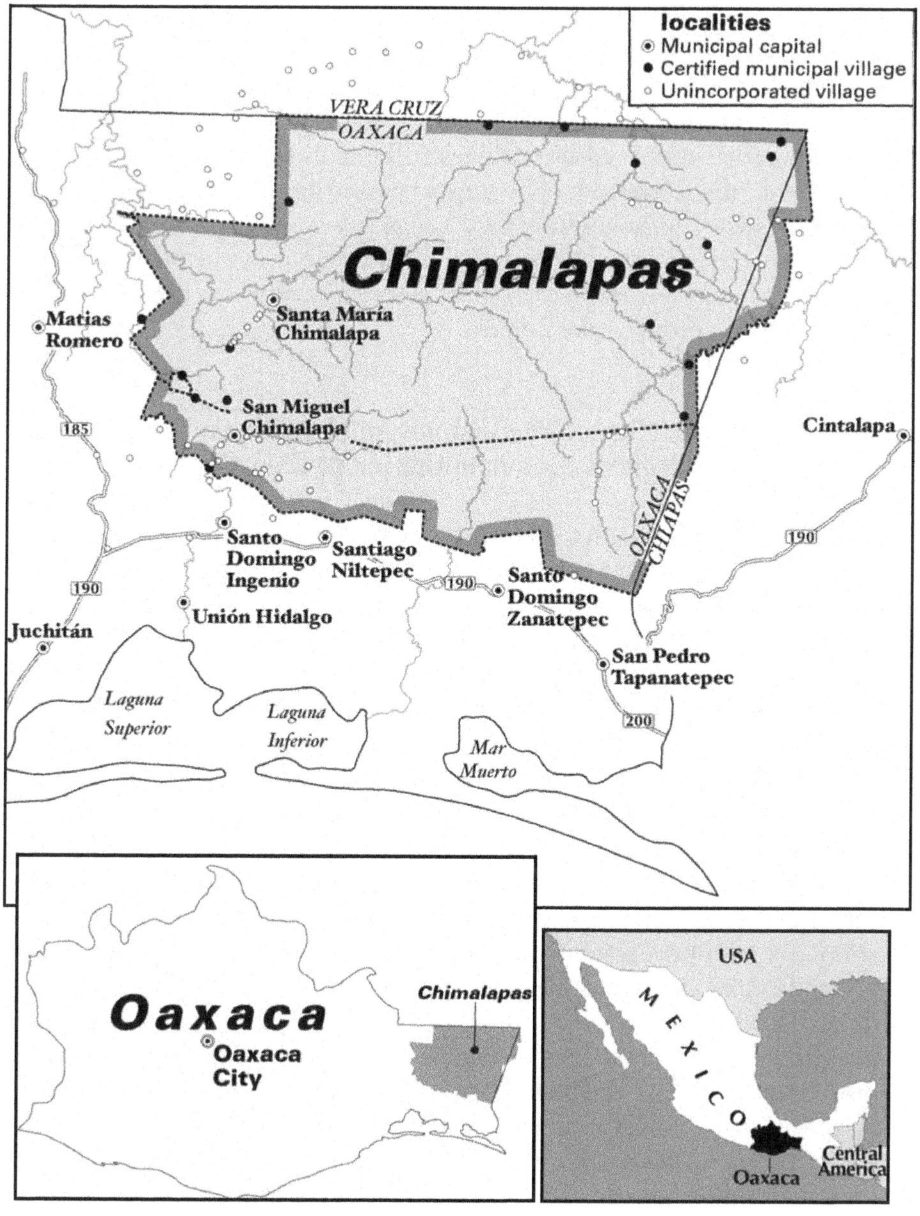


FIGURE 6.4 This image shows three maps. The bottom right side shows the country of Mexico and the location of Oaxaca. The bottom left shows the state of Oaxaca and the location of Los Chimalapas. The top map shows the territory of Los Chimalapas, the headquarter villages of Santa Maria and San Miguel Chimalapas, and the certified municipal buffer villages along the borders of Chiapas and Veracruz (Walker and Walker, 2008).

communication). After 1950, with the exception of one centrally located village, nine new villages were established on the border with Chiapas and along the extreme eastern fringes of the municipalities.

Within the framework of Zoque reverse settler colonialism, it important to understand how Indigenous autonomy emerged across Oaxaca, allowing for the creation of autonomous Indigenous municipalities. The next section explains these legal shifts before learning how power is produced and organized by the Zoques.

## State Government Structure, Customs and Traditions, and Communal Property Communities in Los Chimalapas

The state of Oaxaca is divided into 517 municipalities. Municipalities are part of the three-tiered Mexican political system of the federal, state, and municipal governments, with the municipalities functioning as the basic political unit. The multiplicity of municipalities in Oaxaca has created a fragmented political structure that allowed the PRI political party to create division and control the state. Several groundbreaking changes would decenter the top-down power structure and lead to politically empowering Indigenous communities across the state.

In 1992 Mexico signed the International Labor Organizations Convention 169. A component of this convention required the government to consult with Indigenous communities concerning any development activities affecting their territory—from logging to agriculture or the creation of a biosphere reserve. In Oaxaca, the signing of Convention 169 portended changes to the electoral code, which in 1995 recognized traditional communal electoral practices, called *usos y costumbres*, or practices and customs laws. This recognition of traditional decision-making procedures outside of the political party system was overlaid by sweeping constitutional changes in favor of Indigenous communities and peoples (Oaxaca State Legislature LVI, 1998).

Currently, four out of five municipalities in Oaxaca are governed by *usos y costumbres* and are considered Indigenous municipalities, or *communidades de bienes communales*. This shift in the legal landscape in Oaxaca provided the Indigenous municipalities considerable autonomy and the power over their lands and resources. Reinforcing the spatial practices of the buffer villages was the decision to grant Zoque land the title of *usos y costumbres*, endowing the Zoques the power to self-govern.

The status of *usos y costumbres* normally embodies notions of a common Indigenous history, a knowledge of pre-Hispanic languages, community organization based on kinship ties, and community service and political participation based on one's standing in the community rather than on strict electoral procedures (Oaxaca State Legislature LVI, 1998). *Usos y costumbres* is essentially a codification of the autonomous normative structures that have existed among native peoples in Mexico since precolonial times. Although Indigenous communities continue to be bound by Mexican agrarian law, at the municipal level leaders maintain control over land management issues and the actions of community members primarily through *estatutos,* or written laws, many of which concern conservation and natural resource management (see Walker et al., 2007, in the "Suggested Resources"). For example, the Zoques penned their *estatutos* upon becoming an Indigenous municipality. An important section of the San Miguel *estatuto* called "Community Natural Resource Conservation, Maintenance, and Benefits," dictates the practices of sustainable logging, the protection of endangered species (both flora and fauna), reforestation programs, controlled burning for agricultural production, and ecotourism, among other topics (Estatuto Comunal San Miguel Chimalapa, 2000).

Historically, only male members of the communities are chosen to perform duties for the communal assemblies. This practice has recently changed. Ceyla Cruz Guiterrez, an Indigenous Zoque with a law degree from the Benito Juarez Autonomous University in the city of Oaxaca, and a staunch supporter of Zoque rights who defends Zoque territory, became the first woman to serve in the communal assembly (López, 2023). Her election is an indication that with the professionalization and further education of the Zoque community, tactics partially learned from observing government officials and NGOS, the patriarchal structure of the Indigenous municipalities may be waning. Serving as a member of the communal assemblies involves shifting appointments, meaning elected officials are not elected to a permanent seat, representatives are to leave their positions to allow others to serve. However, often there exist discrepancies and internal power struggles that may lead to an internal group of male members rotating the duties of the Indigenous municipal communal assemblies, thus effectively constructing an internal oligarchy that supposedly governs communally. This practice is reflected to a degree in the Chimalapas.

The Indigenous municipalities of Santa Maria and San Miguel are concentrated legislatively in a governing body, or *comisariado,* consisting of 140 Zoques from the headquarter villages (of which the total population of the Chimalapas is roughly 15,000). Residents from the heterogenous buffer villages are not allowed to participate. The Zoque governing body vote on the agrarian and land tenure decisions for the entire community (personal communication). Through voting mechanisms or strategies this Indigenous minority, which represents roughly 35 percent of the total population of the region, has been able to manipulate space in their political favor and maintain authority over both mestizos and several Indigenous groups. Historically and importantly, women were excluded from the decision-making process, which has changed with Ceyla Cruz Guiterrez's breakthrough election. Nevertheless, women are often mobilized for public protests and meetings with outside groups (Figure 6.5). The presence of women cooking and caring for children can also work to "symbolically underline the powerless position of the Chimas in relation to the state" (Umlas, 1998, p. 185). However, Ceyla Cruz Gutierrez insists that "we are going to strongly promote the heartfelt

FIGURE 6.5 Women providing support to the Zoque officials in the San Miguel Municipal Auditorium. Photo by David M. Walker.

demand that we have heard in the communities, which has to do with respect for the participation of women in community assemblies [or the governing body/*comisariado* meetings] . . . [and] we are going to ensure that women's bodies have respect in the areas of health, education and sports" (Lopez, 2023).

Today the Chimalapas region is spatially and ethnically divided. The two municipal capitals, called *cabeceras* in Spanish, are predominantly Zoque, and the small villages and towns along the border with Chiapas and Veracruz are inhabited by a heterogeneous population consisting of a small Zoque bureaucracy, and Indigenous groups including Zapotecs, Tzotzils, Mixe, Mixtecs, and mestizos, the later group having come to the Chimalapas with logging companies and still residing here. The *cabeceras* are the cultural and political seats of the Zoques. Santa Maria and San Miguel Chimalapas have political jurisdiction over 40 scattered, certified municipal villages and unincorporated villages (personal communication). Although the distinct areas are not strictly defined by linguistic characteristics, residents of the *cabeceras* speak primarily Zoque and Spanish, while the heterogeneous fringe villages are polylingual, with Spanish functioning as the lingua franca. The heterogeneous population that has emerged in the fringe villages is a product of the different waves of colonization in the Chimalapas and a result of the spatial strategies employed by the Zoques to incorporate a variety of populations to prevent loss of territory. The next section explains how the Zoques mobilize and maintain control over their territory.

## Structure in Los Chimalapas

The center of power in Los Chimalapas is reproduced materially and metaphorically through sociospatial strategies that have a history dating to at least the 1800s. The Zoque minority has demonstrated a keen sophistication for identifying and taking advantage of important political, social, and environmental opportunities. The Zoques have astutely used the presence of vital natural (and lucrative) resources and the move toward transparency on the part of state and federal government officials, embedded in larger neoliberal practices, to position themselves as stewards of the Chimalapas. Additionally, the Zoques have used the support of national and international actors represented by artists, activists, and NGO representatives to their advantage to maintain their presence in the Chimalapas and control the resources found in the region (personal communication). Learning from and building upon examples of media manipulation by grassroots movements in states like Chiapas, the Zoques are adept at utilizing Mexican radio, television, and the local and national press to publicize their struggle.

Zoque community members from the headquarter villages organize weekly patrols of the eastern perimeter of the selva, their territory, to ensure that no land invasions occur (Manzo, 2023). The Zoques also place markers delimiting the boundary between the Zoque territory and the border of neighboring Chiapas state for which they provide yearly maintenance. In addition to these strategies, the Zoques have successfully engaged in roadblocks to prevent the incursions by land invaders. When especially threatened, they have relied on extreme measures, such as the apprehension of hostages, including the brother of the governor of Chiapas, who was accused of illegal logging and drug trafficking (Garcia, 1989). The Zoques have also taken on the state, as exemplified in their occupation and subsequent closure of government buildings in Matias Romero, an important legislative town in the Isthmus, in order to protest land invasions.

Another strategy has involved marches on Mexico City, the nation's capital, demanding an audience by the national legislature and the president, which attracts much publicity and media coverage that brings attention to their plight (Manzo, 2023). When thought to be necessary, the Zoques have resorted to violent armed confrontations with powerful state and corporate-backed land invaders.

The Zoques have implemented these strategies to protect their territory for the very reason that the Chimalapas is so rich in natural resources and geopolitically important. The next section explains why their territory is so sought after by NGOS who seek to create a biosphere reserve, by logging companies, and by the federal government with its own development plans for the region.

## Geography, Resources, and Politics in Los Chimalapas

Geopolitically, the area is positioned as a prime focal point for national and multinational companies. The Isthmus is just 90 miles wide and functions as a transportation corridor and as a key site for large-scale development projects. Indeed, many companies have bought land in anticipation of the implementation of the Plan Puebla Panama—a mega-project linking the Pacific Ocean and the Gulf of Mexico across the Tehuantepec Peninsula. PEMEX, the state oil company, in tandem with foreign capital, is finalizing the construction of a trans-Isthmus transportation service from the Laguna de Pajaritos Port (in Coatzacoalcos, Veracruz) to Salina Cruz via a pipeline. The infrastructure project includes petroleum loading and unloading docks on each side of the Isthmus. The Gulf Coast is rich in crude oil. Now the oil can be quickly pumped across the 90-mile Isthmus and processed at the Salina Cruz refineries for west coast domestic consumption and for international distribution. Additionally, the trans-Isthmus transportation service functions as a conduit for exporting natural gas from Mexico and also from the United States and other countries to the Asian market (Ordaz, 2020). This development project reflects how the Panama Canal is becoming obsolete for shipping the most-traded commodity in the world, as the canal's antiquated infrastructure can no longer accommodate huge oil tankers. Transport of goods by the interoceanic route of the Isthmus of Tehuantepec is much faster than crossing the Panama Canal, with week-long delays causing shipping companies monetary losses.

The greatest threat to the Chimalapas is the conversion of forest land for cattle ranching and logging purposes, which claims upward of 250,000 acres per year. The cultivation of marijuana and opium also shifts land-use patterns and harms the environment (Aguirre, 2015; Doane, 2017). As more Central American political refugees flee violence at home and migrate north to the United States, the pressures of having more people moving through forest land also impacts the environment (Leyva-Flores et al., 2019).

From a social perspective, a number of issues characterize daily life; they include the unabated exploitation of the forest; narcotrafficking; human smuggling; the trafficking of rare and endangered species; violent land disputes; and a lack of roads, electricity, schools, hospitals and potable water highlighted earlier. The natural and political attributes of the region have prompted countless alliances and intercessions among diverse groups, including ecologists and NGOs. The next section describes and provides some examples of how the Zoques negotiate these challenges.

## Confrontation and Cooptation in Los Chimalapas

The trajectory of Zoque-directed colonization has been arduous and at times tenuous, as cattle ranchers and lumber merchants, with the support of the Chiapas government, have sent Tzotzil Indians to foment conflict with people in fringe areas. In addition, Indigenous groups from Chiapas along with cattle and timber interests refuse to recognize Zoque communal property rights (Erguiliz, 1995; Escalante and Romero, 1998; Manzo, 2023). These land invasions represent a significant obstacle to the Zoque elite, but those in power have made inroads in resolving agrarian conflict by reconciling with Chiapas *campesinos*. Critically, and somewhat impressively considering the context, the buffer villages are politically controlled by the Zoque-run headquarter villages employing the tactics described earlier, such as patrols and border delimitation.

The project of endowing Indigenous communities with the power to self-govern described earlier coincides with the conscious shifting of identities of the Zoques, where "the emergent political actor in Oaxaca was no longer peasant, but Indigenous and ethnicity more than class became crucial in the mobilization of political identities" (Anaya Muñoz, 2004, p. 427). This identity shift and Indigenous professionalization facilitated the opening of social spaces for the endorsement of Indigenous rights as a means of resolving their own needs and wants and advancing their own political agendas as rational citizens, as long as these demands do not interfere with the capitalist order. The spatial strategies employed by the Zoques have involved relationships with different actors and numerous instances of cooptation on the part of members of the Indigenous group. The logging company Sanchez Monroy, which hired workers from all over Mexico to work in the timber industry in Chimalapas between 1950 and 1970, provides an instructive example.

Upon arriving in the Chimalapas, the Monroy employees illegally formed villages in the eastern part of Zoque territory, and this practice continued unabated until 1967. These models of land tenure allowed the Zoques to effectively organize and create unions with a wide range of other groups, including, in 1977, the workers from Sanchez Monroy. The Zoque elite, with the help of loyal inhabitants in many of the surrounding scattered towns, blocked the roads constructed by the logging company and hijacked their machinery, effectively discouraging Sanchez Monroy from operating in the region. Villages populated by former workers were incorporated and now answer politically to the respective *cabeceras*. The buffer villages function as a barrier to the expansion of logging companies and the intrusion of farmers and loggers from Chiapas. These villages have been the sites of intense armed conflict where a number of inhabitants have lost their lives while defending the Zoque territory (personal communication).

The Zoques reaction to government-development schemes provides an additional example of the Zoques ability to protect their territory. In 1947 the Mexican federal government, under the leadership of President Miguel Aleman Valdés, established the National Colonization Commission with the goal of redistributing the population in the territory of Uxpanapa located in Veracruz. Similar to federal government plans in Brazil in the 1960s and 1970s that reimagined the Amazon forests in the Amazon River basin as an escape valve for densely populated urban centers in the northeast and south of the country, the National Colonization Commission's goal was to exploit and populate what were constructed as "empty areas of the national territory" (de Teresa, 2000, p. 91), often through development programs aimed at making use of natural resources.

Between 1947 and 1960 the National Colonization Commission carried out seven repopulation schemes in relation to hydrological development projects, including the construction of the Miguel Aleman Dam and the Cerro do Oro Dam in Veracruz in the Papaloapan watershed. As described below, the Cerro de Oro project directly impacted the Chimalapas region. The 17-year development project inundated 32 *ejido* communities and required the displacement of some 10,000 *Chinanteco ejidatarios,* or shareholders, who were eventually relocated to "empty" areas of the Mexican territory. The Commission offered four relocation options to the *ejidatarios* on the fringes of the reservoir in the Veracruz municipalities of Tres Valles, and Uxpanampa. The majority of them opted to stay in the area where the reservoir was to be formed, and about one-third chose to relocate to the northern frontier in the municipality of Santa Maria Chimalapas, where they began settling in the late 1970s through 1989 (Aguirre, 2015; Garcia, 1999).

The government-sponsored repopulation scheme was carried out chaotically. Some non-Chinantecos posed as the displaced Indians and received plots in the Chimalapas. Also, some Chinantecos accepted the plots on the border of Chimalapas and Veracruz, only to sell them to non-Chinantecos who were not part of the repopulation plan (deTeresa, 2000). The Zoques in the headquarter village of Santa Maria were alarmed by the usurping of their ancestral lands and proceeded to stop the encroachment of the Chinanatecos. The Zoques ensured this outcome via two strategies.

First, they invited the Chinantecos to become officially part of the Chimalapas territory, thereby placing the Chinantecos villages under the control and jurisdiction of the Zoques at Santa Maria. And second, they invited landless Mixtec peasants living near Sola de Vega in the Oaxaca Valley to form two buffer villages near the Chinanteco communities in order to further prevent invasions and physically guard the jungle's natural resources. As one Zoque official stated, it was seen necessary to form the Mixtec buffer village because: "Well, what we did was to send people there to populate the lands, you know, to protect them so the people from Chiapas wouldn't invade them. Even if the *comisariado* [the Zoque commission that votes on and enforces the written laws of the Indigenous municipality described above] or the state removes them from our lands, in which case they end up with nothing, in the end they always come back. That's why we send people there (along the border of Veracruz and Chiapas) to settle those lands. It is important to protect and preserve *our* natural resources" (personal communication). The Zoque community members viewed the Indigenous Chianteco and Mixtec subsistence farmers as less of a potential threat to the resources of the Zoque jungle than renegade loggers, who were often backed by capitalist and state government interests from Chiapas.

To understand how the Zoques have managed to maintain control over their territory requires an in-depth analysis of how and why they frame themselves as part of the Chimalapas. They see themselves as part of and not separate from the other resources found in the Zoque jungle.

## The Conflation of Nature and Society in Los Chimalapas

Considering the various social and spatial processes at the heart of daily life in the Chimalapas, how is it that the governing minority that struggles to maintain political power over dispersed villages has been able to exercise concrete and symbolic actions that have resulted in control over varied populations both within and outside the buffer villages? Such control is enacted through material and discursive practices that are embedded in a larger framework of particular conceptualizations of space and "nature." A reading of Zoque-written *estatutos*, coupled with conversations with Zoque leaders, reveals a decidedly geographical theoretical point of view, whereby space is not understood as an empty container but is a reflection of social relations defined by cultural idiosyncrasies and networks of power: that is, space, as in the buffer villages along the border of the Chimalapas, is socially produced (Escobar, 2012). The living, changing nature of the territory can never be represented one-dimensionally through government-produced maps, but must be experienced daily with the understanding that space and people are mutually constitutive.

In both the Chimalapas Communal Statute, a book of laws written by Zoque *comuneros* from San Miguel Chimalapa and in an interview with Zoque leaders, the Zoques envision themselves as an extension of the land: "We are Chimalapas and Chimalapas is us—our ancestors are in this land as are we, with the forest, the animals, the trees" (personal communication). Here, the biodiversity of the region is conceived of in terms of territory plus culture. For the Zoques, there is no distinction between space, place, and being. Nature cannot be bounded and separated from the body, as the Zoques comprehensive notion of territory involves enlisting self-representational strategies enveloped in the complex politics of recognition. Ethnicity and the need to exhibit the validity of Indigenous understandings underpin social mobilizing and political demands in the struggle for self-determination and the claiming of Zoque ancestral lands.

The question then becomes self-determination for whom? While conducting research in the municipal capital of San Miguel Chimalapa, I witnessed firsthand the Zoques' use of strategic tactics. I spent an afternoon at the offices of the Indigenous municipality where members of the *comisariado* met with "subjects" or people living in the buffer villages to discuss and settle issues ranging from land disputes, border disagreements, and the allocation of federal and state funds. On that day, the *comisariado* resolved, albeit temporarily, a quarrel among subjects from two certified municipal villages from the eastern Chimalapas regarding their rights to use pasture lands for grazing cows. The subjects from one village were grazing their cows on the communal lands of another village without providing any compensation or sharing the benefits of the livestock in the form of milk or meat. The discussion among the villagers became so agitated—with each member aggressively attacking one another over who had rights to utilize the land—that I feared a violent outburst. Moving to resolve the conflict, the *comisariado* referred to a law in the handbook that had been written by Zoque community members from San Miguel. Angry villagers were shown maps delineating the divisions of the municipally incorporated villages and demarcating the boundaries among the buffer villages and the divisions between Chimalapas and Chiapas. The Zoques had made the maps themselves, with the aid of a local NGO by the name of Maderos de Pueblo, and the law referred to in the handbook referenced the autonomous rules and regulations that govern the entire municipality of San Miguel, all of them written by the Zoque governing body. When the dissatisfied subjects from the fringe villages continued to protest, criticizing the authenticity of the maps and the source of the boundary lines, the head of the governing body stated, "We have the hereditary rights to make decisions over this forest as we have inhabited the forests since time memorable" (personal communication). The dispute, albeit most likely temporarily, halted.

The head of governing body, through this statement, employed a discursive mechanism to position the Zoques as legitimate stewards of the Chimalapas. In human geography the term *discourse* represents the relationship between the constitution of knowledge and truths on one hand and social power relations on the other. Discourse analyses examine how images and stereotypes of spaces are created, how spatial identities are constituted, and how spatial normative political decisions are legitimized by reference to particular norms and values (Gregory et al., 2009). The deployment of discourses that work to position the Zoques in particular ways is inextricable from sociospatial power. Most notably, an "ethnicist" discourse constructs the Zoques as the true inheritors of the land with innate stewardship responsibilities. The elite Zoques place themselves in a superior position through constant reaffirmation of their inherent natural rights to reside in the Chimalapas, while also emphasizing their kindness in *inviting* groups to populate the surrounding areas. Within the specified territory, there is a hierarchy between central populations and periphery populations—those living in the buffer villages are called "subjects" by residents of the headquarter villages. The material reality that the Zoque ruling elite has more access to resources, more consolidated power, their own language, and a higher standard of living relative to people in the fringe villages, is used to legitimize their rule, thereby locating the Zoques as superior to the non-Zoque population. This statement further exemplifies such discourse: "It is that the Chimalapas are the lungs of Mexico, no it is more than that, the Chimalapas are the lungs of the world and it is our responsibility as Zoques to take care of the nature and the biodiversity found in the Chimalapas. That is why we organized and founded the *congregaciones* [buffer villages] in the east of the region [on the border of Chiapas] in order to protect these natural resources that belong to us as Zoques; because we are the inheritors of this land called the Chimalapas" (personal communication). Embedded within these statements is the oral and written diffusion of the Zoques' historical ties to the region. The Chimalapan communities existed prior to the colonial period and formed the first independent Mexican territory. At the intersection of legend and history is the story that the Zoque community ceded to the Spanish crown enormous jars of gold, called "Chimalapas" in the Zoque language, in exchange for recognition of the Zoque territory. Historians have identified this event in the Archives of the Indies in Seville, where, in 1687, the territories are recognized and the municipalities of Santa Maria and San Miguel Chimalapas were formed (Aguirre, 2015; Warman, 2003).

More than 300 years later in 1967, Mexican President Gustavo Diaz Ordaz passed a decree recognizing 594,000 hectares (about 2,293 square miles) as a community based on Zoque identity that would become the municipalities of Santa Maria and San Miguel Chimalapas. Fast forward to the legal decrees in the late 1990s that allowed for a legal shift from government municipalities to *communidades de bienes communales* governed by *usos y costumbres* (Indigenous municipalities governed by practices and customs laws, as described earlier) and organized by *estatutos*, or laws written by the Zoque governing body. The decrees were problematic from its inception for several reasons. First, they were written by people from outside the region with less than meticulous attention to the on-the-ground realities of land occupation. Second, the population that had inhabited the buffer villages was not recognized as part of the communities of Chimalapas, or the Zoques, and were effectively considered squatters and land invaders. And third, *ejido* lands had already been given to various groups of Chiapas peasants within the perimeter of the Zoque jungle, the ethnic and historical territory of the Zoques; these lands were called *communidades de bienes communales de Los Chimalapas*. As a further complication, parts of this same territory had been designated for the creation of agricultural communities, and many of the residents had gained title to the land as private property owners (deTeresa, 2000; Warman, 2003).

This legal landscape, fraught with tension and uncertainty, is a backdrop against which discourse takes on supreme importance for the Zoques. Through the establishment of a deep connection to the land, the governing group of Zoques is transformed into something other than "common citizens." As in many instances in Latin America, the loss of territory is equated with a return to slavery and/or the loss of a livelihood (Escobar, 2012). It is important to point out that ideas of Indigenous peoples as culturally distinct groups with conservationist tendencies represent a complex articulation of history and an engagement with transnational discourses that have sometimes equated the Indian subject with the environment. In Chimalapas, we see a larger pattern emerging, "the effort of Indigenous leaders consciously to enact [ing] and diffuse[ing] the images and discourses of a collective indigeneity produced about them (as conservationists and traditional peoples)" (Valdivia, 2005, p. 292). For the Zoques, the realization of the potential effects of discourse has motivated an informal campaign to obtain community recognition in a national and international context. For example, Maderas del Pueblo del Sureste, A.C., an ecological NGO that worked in the region, was instrumental in creating awareness not only about the biodiversity of the forest, but also about the ongoing struggle of the Indigenous communities, casting them as the "ancestral caretakers" of the territory. *La Jornada*, a widely read Mexican newspaper, published a letter written by the Zoque governing body of Chimalapas, that critiqued the outside influences represented by ecological groups and NGOs as reinforcing "private" discourses in the "public" domain. Dissemination of information through various media outlets worked not only to legitimize land tenure claims, but also became crucial to the reproduction of the material and symbolic worlds of Indigenous subjects.

## Ethnospatial Politics

One of the most striking observations to come out of numerous interviews with Zoque leaders and analyses of documents written by the governing body is that those in leadership positions are involved in contradictory processes of differentiation and homogenization. The Zoques example is reflective of how limited membership in a legally sanctioned community organization can be implicated in the production of difference involving a complex geography of displacement and juxtaposition (Gregory et al., 2009). The strategically located buffer villages are identified in relation to the headquarter villages and in their interaction with the Zoque minority, both with regard to hegemonic maps and in everyday practice.

The spatial strategies also work to homogenize through symbolically revising the kinds of spaces that are produced. Instead of buffer communities, the villages are imagined as bounded heterogeneous Indigenous spaces where natural resources are not abstracted from their cultural surroundings, which pertain to the true Zoques, the stewards and inheritors of the Chimalapas.

The Zoques mobilize their indigeneity and embrace the concept that Indians are true and good stewards of nature and natural resources, which effectively temporarily downplays the differences among the heterogeneous actors in the Chimalapas and assumes unity for the sake of achieving their political goals.

A point of convergence between the Zoque inhabitants revolves around the concept that nature cannot be thought of as separate from culture, representing a distinct view from the nature–culture binary conceptual framework for understanding the world that many NGOs and outside groups embrace, including ecologists who have been active in the region for the past 25 years. By placing questions of Indigenous identity at the center of the conversation, the Zoques are able to recast (if only momentarily) these spaces and represent them in relation to the larger heterogeneous Chimalapas identity through the mobilization of strategic essentialism. The Zoques mobilize their indigeneity and embrace the concept that Indians are true and good stewards of nature and natural resources, which effectively temporarily downplays the differences among the heterogeneous actors in the Chimalapas and assumes unity for the sake of achieving their political goals. The homogenizing effects should not be overemphasized, however, especially when tackling the socially constructed concepts of Indigenous or Indian. The seemingly homogeneous character of the collective (i.e., Zoques as Indigenous, and not peasants, who are part of and not separated from the Chimalapas territory) obscures the dynamic processes of Indigenous self-identification and contestations that take place within communities that constitute the collective.

The Chimalapas provides an example of Indigenous groups implementing global strategies whereby community members are "able to gain political and financial support for their claims from actors with broader institutional reach and at the same time to link their concerns to struggles at the regional and national scales" (Perrault, 2003, p. 97). The processes of community formation and land titling in the Chimalapas involve the spatial restructuring of settlements whereby Zoques are engaged in a form of spatialized politics. The alliances the Zoques have created are based in part on beliefs about communal land, on antigovernment sentiment, and on an ethnicist discourse that mobilizes signifiers such as "Indigenous" and "stewardship." These alignments complicate binaries such as "powerful" versus "powerless" and *mestizo* versus *indio* by chipping away at the already fledgling notion of the local as a site devoid of power. What occurs in the Chimalapas is a counter example to the traditional historical trajectories in which Indigenous peoples are often at the mercy of a pigmentocracy and/or government policy.

## Conclusion

For now the Zoques still maintain control over the ancestral territorial limits of the Chimalapas from the headquarter villages of Santa Maria and San Miguel. Through the discourses, strategies, and practices that have been discussed, they maintain territorial and political control over the buffer villages and do their best to prevent the continuous encroachments of loggers and landless peasants from the neighboring state of Chiapas.

By mobilizing the strategy of conflating culture–nature constructs and by positioning themselves as the embodiment of the Chimalapas, the Zoques have cooperated with and co-opted NGOs into not distinguishing social justice goals from environmental conservation goals in Los Chimalapas. The Zoque demands for the inclusion of social justice goals proved reminiscent of the similar demands made and achieved by the neighboring Zapatistas in Chiapas. Because of the perceived radical demands of social development that the Zoques insist be included in any conservation rubric (i.e., park, biosphere reserve, campesino ecological reserve), no such project has yet to be achieved. International interest in Plan Puebla Panama has been more difficult for the Zoques to contend with. The infrastructure projects have been implemented, and plans to bring development to the region continue.

The successful Zoque strategies of scaling their efforts to control the Chimalapas from the local to the national and beyond have led the federal government to intervene into the territorial disputes among loggers and farmers from Chiapas and the Zoques of the Chimalapas. This intervention ended a 10-year border dispute in which

> The Zoques territorial defense and protection of the natural resources found in los Chimalapas against illegal loggers, cattle ranchers, politicians and Narcos from Chiapas, who attack and plunder indigenous communities in Chimalapas, are the *driving force* that oversees in the territorial border depute between Chiapas and Oaxaca in the Chimalapas Jungle. (Hernández Navarro, 2021)

The Federal Supreme Court issued a resolution, after a 9-year legal process, that 160,000 hectares (or about 395,368 acres) of land in four *ejidos* and 10 municipalities claimed by loggers and farmers in Chiapas belong to Oaxaca. The court decreed that within 30 months of the sentence the legislatures of both states must correct their respective constitutions and laws to establish the border that the court determined. (El Financiero, 2021). Will these *ejidos* and municipalities become future buffer villages administered by the headquarter villages?

Well, apparently not yet. The governor of Chiapas has refused to recognize the court's decision. The Chiapanecos, an Indigenous group who according to the Supreme Court, are now in Oaxaca, in the community of Santa Maria, are siding with the governor of Chiapas. Moreover, *ejido* members from the *ejidos* of Cintalapa and Dr. Belisario Domínguez, a municipality located in the disputed territory, led by Belisario Domínguez's president, Alfonso de Jesús Gómez Mendoza, announced they are not in agreement with the court's decision, and they do not plan to evacuate their villages. President Mendoza petitioned to go before the Interamerican Commission on Human Rights because he feared that "there is a risk that the neighboring communities of Oaxaca [the Zoques] will invade . . ." (El Financiero, 2021) claiming that Zoques have already begun to burn their *milpas*, or agricultural plots. The Zoques, in turn, organized a 4-day roadblock of an important Isthmus transportation artery, which got the attention of officials, who agreed to a meeting with the Zoques in San Miguel Chimalapas. As they did not view more meetings as productive since the Supreme Court had decided that the 160,000 hectares of land in four *ejidos* and 10 municipalities claimed by loggers and farmers in Chiapas belong to the Zoque Indigenous land, the Zoque leaders felt that "It is urgent that the governments recognize us, this process has been delayed, [we] are the ancestral owners of our communal territory, that is why we demand the immediate total annulment of all the presidential resolutions, demarcation certificates and plans of all the *ejidos*, and towns that [the Chiapnecos formed when they] invaded in our communal territory [of] Santa María and San Miguel Chimalapas," insisted governing-body member Vidal López Hernández (Manzo, 2023). When officials did not show up, the Zoques planned a "vehicular march" to the nation's capital where they held an assembly that was posted online.

> the National Committee for the Defense and Conservation of Los Chimalapas invites citizens to a National Assembly in Mexico City to share and update information about the Supreme Court ruling and about the Interoceanic Corridor [the Plan-Puebla Panama pipeline]. The assembly will be held on Sunday, July 2, from 11 a.m. to 3 p.m., in the geodesic auditorium of the Huerto Roma Verde (Calle Jalapa #234, corner with Coahuila, Colonia Roma Sur). The event will be broadcast live on Facebook at *www.facebook.com/Istmo Nuestro* and on YouTube at Los Tejemedios (Manzo, 2023).

These events demonstrate how the Zoques continue to mobilize support from the headquarter villages through road blockades and caravans to the country's capital, Mexico City, where they organized public protests that were broadcast virtually from a borough hall located in an upscale neighborhood of Mexico City, thus demonstrating the Zoque's adeptness at implementing a multiscalar approach to protect their historical land, which is territory they conceptualize as part of them as much as they are part of the selva.

Based on contemporary and historical events, I anticipate that the Zoques will continue to manipulate geographies for political ends. That is to say, these linguistically and ethnically heterogeneous colonized spaces will continue to be characterized by an elite Zoque bureaucratic minority that facilitates hegemonic control of the fringe areas. However, it is important to mention that the sociospatial landscape will become increasingly muddled due to the geopolitical strategic location of the Chimalapas and the array of actors involved in formal and informal politics. In light of the on-the-ground practices in the Chimalapas, it is worth asking whether the activities of this minority elite and subsequent alliances are examples of fundamental changes regarding who and what constitutes a powerful group (especially important given the history of race, class, and ethnicity in Mexico), or if these practices are an extension of the colonizing dynamic and, if not anchored directly in state policy, then either consciously or unconsciously mimicking it?

## DISCUSSION QUESTIONS

1 Considering the various social and spatial processes at the heart of daily life in the Chimalapas, how is it that the governing minority that struggles to maintain political power over dispersed villages has been able to enact both concrete and symbolic policies that have resulted in control over varied populations both within and outside the buffer villages?

2 In Chapter 3 and in this chapter, we see how identity is not fixed, that it can change over time, and, in fact, much like the social production of place and space, identity is always in the process of becoming. This can be seen in how the Zoques' identity shifted from peasant to Indigenous as ethnicity more than class became crucial in the mobilization of their power to protect the Chimalapas and in how Afromexicano NGOS mobilized black politics of identity to be recognized as Afromexicanos (*Afromexicanos*), instead of *pardo* or *moreno* or by their geography as *Costeños*. Indeed, Chapter 3 looks beyond the shifting categories of identity to discuss how identities can be fluid: "It is important to be *pardo*, moreno, or *costeño* and also have solidarity with African descendants from Veracruz and Coahuila at the national level, or Cuba, Columbia, Brazil, and even the United States at the international level under the umbrella of Afromexicano. The Afro, in Afromexicano, links the *Costeños* together with the African diasporic communities from across the Americas. But place and geography matter too." This means that individuals can perform multiple identities according to the spaces they are negotiating. Can you provide any examples of shifting or fluid identities in your community or society (perhaps think about changes in the U.S. Census, or the use of preferred pronouns)? Now, describe (1) in what culture(s) do you participate; (2) to which racial or ethnic group(s) do you identify; (3) what is the basis of your primary cultural identity; and (4) do others readily recognize this basis and identity, why or why not?

3 In Chapter 3, *Identity and Place: The Social Geographies in the Costa Chica de Guerrero and Oaxaca*, and Chapter 6, *Zoque Indigenous Produced Space: Social Geographies of Los Chimalapas*, in the creation of buffer villages and the protection of Los Chimalapas we see how identity is not fixed, that it can change over time, and in fact, much like the social production of place and space, identity is always in the process of becoming. This can be seen in how the Zoques' identity shifted from peasant to Indigenous as ethnicity more than class became crucial in the mobilization of their power to protect the Chimalapas. The chapter discussed how Afromexicano NGOs have mobilized the Black politics of identity to be recognized as Afromexicanos, instead of as *pardo* or *moreno*—meaning Black—or by their geography as *Costeños*. Indeed, the chapter looks beyond the shifting categories of identity in discussing how

identities can be fluid: "It is important to be *pardo*, *moreno*, or *costeño*, and also have solidarity with African descendants from Veracruz and Coahuila at the national level, or with Cuba, Columbia, Brazil, and even the United States at the international scale under the umbrella of Afromexicano. The Afro, in Afromexicano, links the *Costeños* with the African diasporic communities from across the Americas. But place and geography matter too," meaning that individuals can perform multiple identities according to the spaces they are negotiating.

4 Describe how the Zoque *comisariado* members became professionalized in the language and practices of conservation and preservation.

5 Describe three spatial strategies the Zoque employ to protect the Chimalapas.

6 What tactics have the Zoques have used to protest against the state and bring national attention to their plight described in the section Ethno-Spatial Politics and in the Conclusion?

7 Explain how the Zoque leadership is involved in contradictory processes of differentiation and homogenization of Zoque identity and the identity of residents residing in the buffer villages.

8 What does the term discourse mean in human geography and how did the Zoques implement discursive practices to position themselves as legitimate stewards of los Chimalapas.

## SUGGESTED RESOURCES

To learn more about the professionalization of Indigenous groups and NGOs in Los Chimalapas, see:

Walker, D. M., Jones, J. P., Roberts, S. M., and Frohling, O. R. (2007). When participation meets empowerment: The WWF and the politics of invitation in the Chimalapas, Mexico. *Annals of the Association of American Geographers*, 97(2), 423–444.

## REFERENCES

Aguirre, M. Á. G. (2015). *Chimalapas: La defensa del territorio y de los bienes naturales como un factor de identidad indígena Centro de Estudios para el Cambio en el Campo Mexicano.*

Anaya Muñoz, A. (2004). Explaining the politics of recognition of ethnic diversity and indigenous peoples' rights in Oaxaca, Mexico. *Bulletin of Latin American Research, 23*(4), 414–433.

De Teresa, A. P. (2000). *Los Vaivenes de la Selva: El Proceso de Reconstitución del Territorio Zoque de los Chimalapas*. Oaxaca, Mexico: Secretaría de Medio Ambiente, Recursos Naturales y Pesca.

Doane, M. (2017). *Stealing shining rivers: Agrarian conflict, market logic, and conservation in a Mexican forest*. Tucson: University of Arizona Press.

El Financiero. (2021). Los Chimalapas pertenecen a Oaxaca y no a Chiapas, sentencia la Suprema Corte: El órgano de justicia dio la razón a la controversia constitucional entablada contra la anexión de 160 mil hectáreas por parte de Chiapas. Retrieved from *www.elfinanciero.com.mx/nacional/2021/11/11/los-chimalapas-pertenecen-a-oaxaca-sentencia-la-suprema-corte*.

Erguiliz, A. (1995). Los comuneros de Santa Maria Chimalapa, Oaxaca y la deforestacion de su selva. In *De bosques y gente: Aspectos sociales de la desforestacion en America Latina*. Cuernavaca: Universidad Nacional Autonoma Metropolitana-Iztapalapa, Mexico.

Escalante, J. M., and Romero, F. (1998). Recorrido de la region del Valle de Uxpanapa. Universidad Nacional Autonoma Metropolitana-Iztapalapa, Mexico.

Escobar, A. (1995). *Encountering development*. New Jersey: Princeton University Press.

Escobar, A. (2012). *Encountering development: The making and unmaking of the third world*. Princeton, NJ: Princeton University Press.

Estatuto Comunal San Miguel Chimalapa. (2000).

Garcia, M. A. (1989). The Chimalapas: In defense of a strategic region. *The other side of Mexico, 12*, 6–8.

Garcia, M. A. (1999, July 8). La gran defensa. La Jornada. Available at *http://jornada.com.mx/1999/07/14/oja-defensa.html*.

Gimenez, G. (1996). Territorio y Cultura . *Estudios sobre las Culturas Contemporáneas, 2*(4), 9–30.

Gómez, E. (2004). Fox entrega la tierras a Santa Maria Chimalapa. *America Latina en movimiento*. Retrieved June 22, 2023, from *http://alainet.org/es/active/5806*.

Gregory, D., Johnson, R., Pratt, G., Watts, M., and Whatmore, S. (2009). *The dictionary of human geography*. New York: Wiley Blackwell.

Hernández Navarro, L. (2021, November 16). Chimalapas, Triunfo Chima. *La Jornada*. Retrieved from *https://www.jornada.com.mx/2021/11/16/opinion/017a1pol*.

Instituto Nacional de Estadística y Geografía (INEGI). (2020). Lengua Indígena. Retrieved June 18, 2024, from *http://inegi.org.mx/temas/lengua/#Tabulados*.

Leyva-Flores, R., Infante, C., Gutierrez, J. P., Quintino-Perez, F.,

Gómez-Saldivar, M., and Torres-Robles, C. (2019). Migrants in transit through Mexico to the US: Experiences with violence and related factors, 2009–2015. *PLOS One, 14*(8). Retrieved from *https://doi.org/10.1371/journal.pone.0220775.*

López, A. (2023, January 10). Ceyla Cruz, defensora del territorio zoque y la primera mujer en gobernar Los Chimalapas, en Oaxaca. *El Universal.* Retrieved January 10, 2023, from *https://oaxaca.eluniversal.com.mx/sociedad/ceyla-cruz-defensora-del-territorio-zoque-y-la-primera-mujer-en-gobernar-en-los-chimalapas.*

Manzo, D. (2023). Towns of Chimalapas undertake a motorized campaign to defend their forests.

Oaxaca State Legislature LVI. (1998). Derechos Indigenas en la Legislacion Oaxaquena. Cuadernos de la Comision Permanente de Asuntos Indígenas, no. 3.

Ordaz, (2020). *Gobierno federal inyectará 9 mil 698 mdp en puerto de Salina Cruz.* Retrieved June 22, 2020, from *http://milenio.com/negocios/sct-inyectara-9-698-mdp-puerto-salina-cruz.*

Perrault, T. (2003). Making space: Community organization, agrarian change, and the politics of scale in the Ecuadorian Amazon. *Latin American Perspectives, 30*(1), 96–121.

Stanton, K. (1992–1993). Stewards of the Chimalapas, *Earth Island Journal,* 8(1), 10.

Umlas, E. (1998). Environmental networking in Mexico: The Comité Nacional para la fefensa de los Chimalapas. *Latin American Research Review, 33*(3),161–189.

Valdivia, G. (2005). On indigeneity, change, and representation in the Northeastern Ecuadorian Amazon. *Environment and Planning A, 37,* 285–303.

Walker, D. M. and Walker, M. A. (2008). Power, identity and the production of buffer villages in the "second most remote region in all of Mexico." *Antipode, 40*(1), 155–177.

Warman, A. (2003). *Los indios mexicanos en el umbral del milenio.* México: Fondo de Cultura Económica.

WWF. (2001). *Chimalapas: La ultima oportunidad.* Mexico City: World Wildlife Fund and SEMARAP.

WWF-Mexico. (2005). *Bosques Mexicanos: Selva Zoque.* Retrieved from *http://www.wwf.org.mx.*

Zoque indigenous communities denounce government negligence in protecting a biodiversity hotspot. *Activism, Forest, Mexico.*

# Index

*Note.* Italics in a page number indicates a figure or a table.

## D



## E



## F

## G



## H

## I

## Q



## R

## T

## Z

# About the Author

**David M. Walker, PhD,** is Professor of Geography at the College of Southern Maryland. His international research centers on neoliberal urban governance, nongovernmental organizations, economic policies, and urban change in Oaxaca, Mexico City, and Tijuana. In the United States, Dr. Walker analyzes immigrant urban revitalization of first-tier suburban neighborhoods, contrasted with formal government revitalization planning practices. He is a lifelong traveler who has lived for long periods in Mexico and Spain.